# A NATIONAL ELECTRICAL CODE STUDY REFERENCE

## BASED ON THE 2008 NEC

# A NATIONAL ELECTRICAL CODE STUDY REFERENCE

## BASED ON THE 2008 NEC

Alvin J. Walker

# A NATIONAL ELECTRICAL CODE STUDY REFERENCE

## BASED ON THE 2008 NEC

2009 © Alvin J. Walker

Portions of this material are reproduced from NFPA 70® - 2007 National Electrical Code®,
copyright © 2007, National Fire Protection Association. All rights reserved.

Cover and interior design by Ted Ruybal.

ISBN 13: 978-0-9822975-1-3
ISBN 10: 0-9822975-1-3
LCCN 2009932306
First Edition
1 2 3 4 5 6 7 8 9 10

This is a Wisdom House product, published by Wisdom House Books.
For information, please contact:
Wisdom House Books
15455 Dallas Parkway, Suite 600
Dallas, Texas 75001
Tel. 1-972-764-3222
www.wisdomhousebooks.com

# TABLE OF CONTENTS

*Number in parenthesis indicates number of questions per NEC reference. Total questions: 1385

# CHAPTER 4–EQUIPMENT FOR GENERAL USE . . . . . . . . . . . . . . . . 88

# CHAPTER 5–SPECIAL OCCUPANCIES . . . . . . . . . . . . . . . . . . 129

# ACKNOWLEDGMENTS

I can do all things through Christ which strengtheneth me (Philippians 4:13). For certain, I know that without Christ in my life and his infinite strength, this book would not be possible. The way hasn't been easy and the obstacles have been many, but, by the grace of God, this journey has come to the first of many milestones. And for this I say, "Lord, I thank thee—great is your faithfulness."

A special thanks to my loving mother and late father, who instilled things in me that I have only now come to cherish and appreciate. Thank you to my sisters and brother and their spouses for their love and support and to my late first-grade teacher for her prayers and encouragement.

To my publisher for believing in me and giving me a chance and to my editor for her tireless efforts and attention to details—many, many thanks. Thanks to the proofreader and designers of this book, whom I have never met. And last, to my Christian family and to all of my students and well-wishers who have crossed my path in current and past years . . . thank you.

# PREFACE

"Stop," "Yield," "25 miles per hour," "School Zone," "Slippery When Wet," "Crew Working Ahead." All of these are recognized as signs of precaution and as summonses to take heed. The National Electrical Code also provides a significant precaution in alerting the user to the fact that the codebook is not for untrained persons.

In other words, the National Electrical Code is not a book that one can just pick up, start reading, and immediately gain total knowledge of the code the first time. Let it be understood that the codebook is a massive combination of individual topics, subjects, and electrical situations that are subject to change at any given moment. Having a sound background in such topics as electrical theory and mathematics is certainly needed to begin digesting the abundance of information provided in the code. The electrical terms and terminology alone will require the user to learn to speak another language—code language, that is. Often enough, trying to decipher rules and instructions in the codebook to perform an electrical calculation is quite similar to those dreaded mathematical word problems we all disliked. Problem-solving techniques and interpretation of the code are not mastered easily. However, both are needed to ensure that electrical installations and repairs are performed safely and without potential hazards to both people and equipment.

During my many years of teaching the code and other related electrical subjects, I have always placed myself in the position of the students. I know how frustrating it is to want to ask questions that you don't know how to ask. I also know how difficult it is to read information over and over, only to have found myself going nowhere. For those reasons and others, I was prompted to prepare this book. My sole intentions were to assemble it in a concise, yet simple, manner that would provide instructions to the user without the actual presence of an instructor.

Because I think so highly of the electrical profession, I try always to inspire students to think likewise of this profession. It is a profession in which you are required to think—and at most times think quickly and with sound judgment. Learning the code is an ongoing lesson, and it never ends for anyone.

Study and learn the code, not just to pass an examination but to have knowledge of how to apply the given rules and applications. Even after getting a license, you will still have need of the code. I have seen students or individuals with YEARS of hands-on and field experience who have no knowledge or concepts of the code—yet they want to get their licenses.

To sum it up, I'll use a phrase I once heard on a radio station: "Constant review is a student's glue." Think about it. There is no quick fix.

# INTRODUCTION TO THE
# NATIONAL ELECTRICAL CODE (NEC)

## A. History and Development of the National Electrical Code (p. 1 of the 2008 NEC)

The National Electrical Code (NEC) is Volume 70 (NFPA–70) of the National Fire Protection Association.

## B. Purpose [NEC 90.1(A)]

The purpose of the NEC is for practical safeguarding of persons and property from hazards arising from the use of electricity.

## C. Intention [NEC 90.1(C)]

The NEC is not intended as a design specification or an instruction manual for untrained persons.

## D. General layout of the NEC

In spite of the difficulties one might encounter in interpreting and understanding the NEC, it is for the most part well structured and serves its purpose. However, to overcome the adversities associated with having to consult the code for information and real-world applications, the first step is to become familiar with the general layout of the code. Without such knowledge, the only guarantees are continued frustration and a lack of appreciation for using the codebook.

To begin, consider the overall format of the NEC:

**1. Contents.** The contents serves as a general guideline for locating and referencing code information.

**2. Chapters.** The NEC is composed of nine chapters. The first eight chapters consist of articles that include sections, parts, and tables, and the final chapter consists only of tables and annexes.

Chapter 1 contains a glossary of essential definitions and general requirements for electrical installations.

Chapters 2 through 4 contain general rules pertaining to wiring installations and use of equipment.

Chapters 5 through 7 contain rules that require specific wiring and installation guidelines for certain occupancies (hazardous locations, mobile homes), equipment (electric signs, fire pumps), and special conditions (emergency systems, fire alarm systems).

Chapter 8 contains guidelines and information pertaining to communication systems (radio, television, and CATV systems).

Chapter 9 exclusively consists of tables and annexes that provides useful information and data pertaining to raceway and conductors. Multiple examples of various type load calculations and other non-enforceable NEC references are also included.

**3. Articles.** The 2008 NEC contains 144 articles that cover specific subjects such as definitions, branch circuits, services, grounding and bonding, auxiliary gutters, appliances, motors, transformers, X-ray equipment, swimming pools, optical fiber cables, etc.

**4. Parts.** Articles that contain an excessive amount of information are divided into parts. Parts are used to individualize the major topics of an article. For example, refer to Article 430 in the contents. Article 430 is broken up into 14 parts. Each part of an article is preceded by a Roman numeral (e.g., II. Motor Circuit Conductors).

**5. Sections.** Sections identify the actual NEC rule(s) or reference(s). For example, in section 220.82(B)(2), the first number, 2 (underlined), identifies the chapter (Chapter 2). When combined with the remaining two numbers before the decimal point, 2 and 0, this indicates the article (Article 220). The number after the decimal point, 82, indicates the section of the code referenced, that is, section 82 of Article 220. The capital letter in parentheses ( ) after the section, (B), identifies a subsection of the section. The number in parentheses ( ) after the subsection, (2), identifies one of the mandatory requirements that are associated with that subsection (at times, the numbers in parentheses may identify certain conditions that are linked to a subsection). Therefore, section 220.82(B)(2) is the rule referenced for determining the specific load requirements for 20-ampere small appliance and laundry branch circuits that will contribute to the sizing of the service load for a single dwelling when using the optional calculation method.

To go a little further, consider requirement (3) to subsection 220.82(B). The letters a.–d. identify specific items or details that are relative to the requirement (3).

**6. Tables.** Tables are provided where numerical information and data are needed to determine specific values, ratings, or allowances pertaining to wiring methods, wiring installations, calculations, equipment, etc. Most tables found in Chapters 1 through 9 are associated with a corresponding section of the code.

**7. Rules.** The NEC contains directives that govern the installation of electrical, signaling, and communication materials and equipment. The NEC recognizes two types of rules.

The first type is the mandatory rule as defined in NEC 90.5(A). Mandatory rules require total compliance. They are identified by the terms "shall" or "shall not." Examples of mandatory rules are NEC 200.9, 210.21(A), 215.2(A)(2), and 220.14.

The second type is the permissive rule as defined in NEC 90.5(B). Permissive rules allow discretionary compliance, where optional or alternative methods are permitted. They are identified by the terms "shall be permitted" or "shall not be required." Examples of permissive rules are NEC 210.4(A), 220.52(B), 225.7(D), and 240.4(B).

**8. Exceptions.** Exceptions provide alternatives to specific rules. They are distinctively written in italics. Consider the following:

Example (the rule): NEC 210.3 states that the ratings for other than individual branch circuits shall be 15, 20, 30, 40, and 50 amperes.

Statement: Multioutlet branch circuits shall not exceed a 50 amperes rating.

Answer: The statement is true based on the rule. Only individual (separate) branch circuits are allowed to have a rating of 50 amperes or higher.

Statement: A No. 10 conductor having an ampacity of 30 amperes supplies a branch circuit that is protected by a 20-ampere circuit breaker. The branch circuit is rated for 30 amperes.

Answer: The statement is false based on the rule. The rating of a branch circuit is based on the ampere rating or setting of the specified overcurrent device. Therefore, the branch circuit is rated for 20 amperes.

*Example (the exception to the rule): The exception to NEC 210.3 states that a multioutlet branch circuit greater than 50 amperes is permitted to supply nonlighting outlet loads on industrial premises where conditions of maintenance and supervision ensure that only qualified persons service the equipment.*

Statement: Two 100 amps multioutlet branch circuits supplying several single receptacle outlets in a steel manufacturing plant are permitted as long as the plant is maintained and supervised by individuals who are qualified to service these type circuits.

Answer: The statement is true based on the condition of the exception to the rule.

**9. Fine Print Notes (FPN).** The FPN identifies explanatory material. Fine print notes are only used to provide information and are not enforceable requirements of the code. Examples of FPNs include those listed in NEC 110.5, 200.7(C)(3), 250.116(3), and 406.9(E).

**10. Definitions.** Part I of Article 100 provides the definition of terms that are used in two or more articles of the NEC. Terms that are common to an article or a particular part of an article of the NEC only apply to that particular reference. Terms that are common to an article are, for the most part, found in the second section (indicated by ".2") of that particular article (e.g., the definitions found in NEC 240.2, 250.2, 366.2, and 440.2). Terms that are common to a particular part of an article are listed according to applicable sections (e.g., Part IX of Article 424). The definitions listed in NEC 424.91 are only applicable to Part IX.

Part II of Article 100 provides the definition of terms specifically covering installations and equipment operating over 600 volts, nominal.

**11. Changes, Deletions, Extracted Text.** Any information listed in the current 2008 NEC that differs from the 2005 NEC is identified as follows:

Changes. Prior to the 2008 NEC, a vertical line or bar ( | ) appearing in the margins of the NEC was used exclusively to identify a section or sections that had been changed (revised). In the 2008 NEC, changes other than editorial are highlighted with gray shading within the applicable section. Vertical lines or bars are used now only to identify large blocks of changed or new text and for new tables and changed or new figures.

Deletions. A bullet (•) appearing in the margins of the NEC identifies an area in which contents have been deleted (removed). A deletion does not mean the deleted contents have been totally removed from the NEC, as they could possibly be located elsewhere in the NEC.

Extracted Text. Prior to the 2002 NEC, a superscript ($^x$) was used to identify material extracted from other NFPA documents. Such extracted (taken from another source) material now appears between brackets ([ ]) in the NEC. For example, 517.34 was extracted from NFPA 99, Section 4-4.2.2.3.2.

## E. Using the National Electrical Code

As an electrical professional, you will be required to reference code information frequently; however, for some this may prove to be quite difficult because understanding the code and knowing how to use the code require study, experience, continuing education, and other required means to become familiar with the contents and material found in the code. The information found in the contents and index of the code only provides limited means for those individuals totally relying upon such resources for locating and referencing needed information.

Becoming knowledgeable about the contents and layout of the code is extremely important and will only come with dedication, personal sacrifice, and perseverance.

Now that the basic structure and fundamentals of the NEC have been discussed, let's focus on selected articles in Chapter 3, which covers cable and raceway (metal and nonmetal conduit, tubing, etc.); see Article 100 for definitions. These articles were selected because they are all formatted the same and are quite simple to obtain information from once understood. Therefore, to gain a working knowledge of how they are used, let's review all applicable sections.

Articles 320–340 cover electrical cable. The intent and purpose of each section is briefly described. Selected sections are only used where applicable (in other words, certain articles will contain a section .6, for example, and certain ones will not).

Section .1: Scope (describes contents of the article)

Section .2: Definition (terms that are only applicable to article of interest)

Section .6: Listing Requirements or Listed (listed material and equipment; see Article 100 for definition of listed)

Section .10: Uses Permitted (how cable is permitted to be used)

Section .12: Uses Not Permitted (how cable is not permitted to be used)

Section .15: Exposed Work (installation of cable when exposed)

Section .17: Through or Parallel to Framing Members (how cable is to be protected based on specific routing)

*Section .18: Crossings (crossings of more than two Type FCC cable runs)

Section .23: In Accessible Attics (how cable to be installed in accessible attics and roof spaces)

Section .24: Bending Radius (required bending radius of cable)

Section .26: Bends (maximum number of cable bends permitted)

Section .30: Securing and Supporting (how cable is secured and supported)

Section .40: Boxes and Fittings (box and fitting requirements for cable terminations)

*Section .41: Floor Coverings (floor covering requirements for floor-mounted Type FCC cable)

*Section .42: Devices (receptacles, receptacle enclosures, and self-contained devices)

*Section .56: Splices and Taps (splices and taps wiring requirements for FCC cable systems)

*Section .60: Grounding (grounding requirements for Type FCC cable systems)

Section .80: Ampacity (cable conductors' ampacity determinants)

Section .100: Construction (cable's physical makeup)

*Section .101: Corrosion Resistance (Type FCC cable protection against corrosion)

Section .104: Conductors (cable's manufacturing requirements)

Section .108: Equipment Grounding Conductor (cable's equipment grounding conductor requirements)

Section .112: Insulation (cable conductor's insulation and types)

Section .116 - Sheath (overall outer covering of cable); Exception: Article 326, Section .116 - Conduit (suitable for use with natural gas rated pipe)

Section .120: Marking (required cable markings)

*Section only applicable to Type FCC cable

The purpose for displaying all applicable sections is to familiarize the user further with the format and structure of such articles to eliminate timeless effort in attempting to gather information. In other words, if there is a need to know how each cable type is permitted to be used, you would automatically go to section .10 of each applicable article for such information. If there is a need to know the bending radius or the number of bends a cable is limited to, you would go to sections .24 and .26, respectively. Get the point? Now that you know how these articles are formatted, finding answers and information quickly is possible, but this only comes with studying and learning the codebook as a whole.

Now let's take a look at Articles 342–362.

Articles 342–362 cover raceways. The intent and purpose of each section is briefly described. Selected sections are only used where applicable (in other words, certain articles will contain a section .14, for example, and some will not).

Section .1: Scope (describes contents of the article)

Section .2: Definition (terms that are only applicable to article of interest)

Section .6: Listing Requirements

Section .10: Uses Permitted (how raceway is permitted to be used)

Section .12: Uses Not Permitted (how raceway is not permitted to be used)

*Section .14: Dissimilar Metals (example: steel-aluminum)

Section .20: Size (manufactured sizes of raceway: minimum and maximum)

Section .22: Number of Conductors (number of conductors permitted in raceway)

Section .24: Bends—How Made (internal diameter maintained, required bending radius)

Section .26: Bends—Number in One Run (maximum number of bends permitted)

Section .28: Reaming and Threading (Trimming) (finished and threading requirements)

Section .30: Securing and Supporting (how the cables are secured and supported)

**Section .40: Boxes and Fittings (fitting requirements for raceway terminations)

Section .42: Couplings and Connectors (raceway's use of couplings and connectors)

***Section .44: Expansion Fittings (expansion fittings for PVC conduit)

Section .46: Bushings (where required with raceway)

Section .48: Joints (requirements for joining raceway between related raceway[s], boxes and fittings)

Section .56: Splices and Taps (splices and taps wiring requirements where used with raceway)

Section .60: Grounding and Bonding (raceways allowed to be used as equipment grounding conductors)

Section .100: Construction (raceway's physical makeup)

Section .120: Marking (manufacturers required raceway markings)

*Section .130: Standard Lengths (standard length requirements for manufacturers)

*Section only applicable to intermediate and rigid metal conduit

**Section only applicable to flexible metallic Tubing, Type FMT

*** Section only applicable to Types PVC and RTRC nonmetallic conduit

As you can see, the formatting and structure is practically the same (with the exception of section topics) and so is the approach for finding answers and information promptly as applied to these articles. Now, if a question or needed information arises pertaining to raceway sizes and grounding, simply refer to sections .20 and .60 of the applicable article.

**F. Practical Exercise.** To conclude the introduction, the following exercises are included to assist the user further in initiating or enhancing his or her understanding of the code.

Exercise 1: Using the code's contents or index, locate the ampacity of a No. 12 THWN copper conductor.

Referring to either the contents or the index will not provide a direct reference for locating the ampacity of a given conductor. Of the two, the index provides the best source, but without having a working knowledge of the code, gathering such information can be quite difficult. Let's begin our research by looking up the word ampacity in the index. The closest term found is ampacities, and even with this listed reference, multiple resources are listed. Beginning with the references provided, let's take a look at NEC 310.15 and selected tables 310.16–310.21. NEC 310.15 provides a great deal of information pertaining to ampacities for conductors and references the same tables per NEC 310.15(B) as found in the index. So now, let's reference the tables. Of the six tables, three can be eliminated because they don't list THWN copper conductors. Focusing on Tables 310.16, .17, and .20, Table 310.20 can also be eliminated because it does not include No. 12 conductors. Now, focusing mainly on the two remaining tables, Tables 310.16 and 17, the ampacity of a No. 12 THWN copper conductor is found to be 25 amps per Table 310.16 and 35 amps per Table 310.17. Because Exercise 1 did not state the operating conditions of the No. 12 conductor per table headings, this is about as far as the question can be answered.

Looking back at the other references found in the index, Tables 310.67–310.86, NEC 366.23, and Tables B.310.1–B.310.10 (which are found in Annex B), additional answers based on such a generic question can also be found in some of these references. Tables 310.67–310.86 and Tables B.310.5–B.310.10 can be eliminated because they don't include No. 12 copper conductors. NEC 366.23 is unrelated and can be eliminated for obvious reasons, which leaves Tables B.310.1 and .3. Although insulation types are not listed in Table B.310.3, it still can be used as a reference because, according to Table 310.13A, all THWN insulation has a 75°C rating. However, if you take a closer look, Tables B.310.1 and .3 provide the ampacities for multiconductor cables and not single conductors (a No. 12 THWN copper conductor is referenced in the question). As a result, the ampacities listed in Tables B.310.1 and .3 also cannot be used to provide an answer to the question.

Exercise 2: What is the ampacity of the No. 12 conductor if installed in a raceway with four current-carrying conductors?

This exercise provides more specific operating conditions. Of all the tables referenced in Exercise 1, only Table 310.16 is applicable, although the ampacities listed in the table only apply to three current-carrying conductors in raceway, cable, or directly buried. Therefore, to determine the ampacity of four current-carrying conductors, the listed ampacity (25A) per Table 310.16 cannot be used because it applies only when three current-carrying conductors or fewer are involved (see heading of Table 310.16). Because we are only having to focus on Table 310.16, there has to be some other references that provide instructions for dealing with such a matter. Because NEC 310.15 is solely based on ampacities for conductors, let's scan through this section with hopes of finding needed information. According to NEC 310.15(B)(2), when the number of current-carrying conductors in a raceway or cable exceeds three, the allowable ampacity of each conductor shall be reduced, as shown in Table 310.15(B)(2)(a). Referring to Table 310.15(B)(2)(a), when the number of current-carrying conductors ranges between four and six, the ampacity of a conductor has to be adjusted by 80 percent (.80). Therefore, the No. 12 THWN copper conductor, which has an ampacity of 25 amps, has to be reduced by 80 percent. Therefore,

$$25A \times .80 = 20A$$

With these results, the answer to Exercise 2 is 20 amps. If an ambient temperature was provided in the question, the corrections factors to Table 310.16, along with the provisions of NEC 310.15(B)(2)(c) if applicable, must be applied.

Exercise 3: Determine the maximum size circuit breaker or fuse that can be used to protect the No. 12 conductor.

Once again, of the two available sources of information, the index proves to be the better one. Now there are three terms included in the question that prove to be key factors. According to the index, these terms reference the use of Article 240. Article 240 provides the general requirements for overcurrent protection and overcurrent protective devices. Because circuit breakers and fuses are overcurrent protective devices, this article appears to be a good place to begin looking for an answer to Exercise 3. Upon moving away from the scope of Article 240, the term "tap conductors," as found in NEC 240.2, provides a section that could lead to an answer, as it references the use of NEC 240.4. After glancing through NEC 240.4, section 240.4(D) was found to be most helpful. Subsection 240.4(D)(5) specifically states that an overcurrent protective device (be it circuit breaker or fuse) protecting a 12 AWG copper conductor cannot exceed 20 amperes. Therefore, the maximum size circuit breaker or fuse that can be used to protect the No. 12 conductor for this application is a 20-ampere device.

Again, let me place added emphasis to the phrase "for this application." Most users think that protecting a No. 12 copper conductor with a 20-ampere device along with protecting a No. 14 copper conductor with a 15-ampere device and a No. 10 copper conductor with a 30-ampere device holds true for all applications, and that simply isn't true. This is why NEC 240.4(G) references the use of Table 240.4(G) for protecting specific conductors, such as conductors being used for motors, welders, etc.

Exercise 4: Refer to Articles 320–362. What does the formatting of each article have in common?

Exercise 4 has already been addressed. See Section E.

Exercise 5: How many circular mils are there in a No. 4 conductor?

First, one has to be familiar with the term "circular mils" to begin addressing Exercise 5. A mil is 1/1000 (.001) inch, simply expressed as one thousandth of an inch. A circular mil, which is also abbreviated cm, cmils, or CM, is the area of a circle that is .001 inch or 1 mil in diameter. A circle is referenced because conductors are circular figures. Therefore, a conductor that has a .001 inch diameter has a diameter of 1 mil and a cross-sectional area of 1 circular mil.

Now that a circular mil has been defined, where do you go from here? Chances are if you are not familiar with or knowledgeable of the code, finding an answer will be challenging. To make a long story short, there is only one place in the code that lists the circular mils of a conductor, and chances of finding it without having a sound understanding of the code are slim. Table 8 of Chapter 9 provides information as to the properties of a conductor; in this case, a conductor's circular mils are a property of the conductor. As you can see, Table 8 is in a remote location, and the probability of even finding it without a great deal of effort or assistance is small. If the index were used, the only reference provided would be listed under the term "conductors," and once again, where would you go from there? Again, if it were not understood that a circular mil is a property of a conductor, the referenced information ("Properties of," Chapter 9, Table 8) as pertaining to conductors would prove to be meaningless. Get the point? This is why a thorough knowledge of the code is crucial.

The answer to Exercise 5 can now be obtained by way of the information listed in the second column of Table 8, which shows there are 41,740 circular mils in a No. 4 conductor.

Exercise 6: How many small appliance circuits are required in a dwelling?

Before getting started, if the user does not have an understanding of what a small appliance circuit or dwelling is, finding an answer will again prove to be difficult.

After referring to the contents and index, the term "small appliance circuit" does not appear at either location. However, the term "appliances" does appear in the index, and listed under this term is the word "small." Based on these findings, two references are listed: NEC 210.52(B) and 550.12(B). Without having to consult both references, NEC 210.52(B) indirectly states that two small appliance circuits are needed in a dwelling, but it also provides a direct reference, NEC 210.11(C)(1), which clearly requires the installation of at least two small appliance circuits in a dwelling.

Hopefully, these exercises, along with all other information provided in the introduction, have made you aware of the need to learn the code. The uncertainties of depending solely upon the contents or index should be clearly recognized. Although the task of learning the code may seem too difficult, it can be done. However, you are the deciding factor. Just remember the old saying: "Where there is a will, there is a way." Godspeed!

# PRACTICAL STUDY QUESTIONS

Before getting started with the practical questions, let's perform an overall review of the code by aligning corresponding articles with the contents of each related chapter. By using this approach, combined with that which was learned in the introduction, the intentions are only to prepare and provide further insight to the users before proceeding with the practical questions.

The 2008 National Electrical Code, like all previous editions, begins with an introduction, which is recognized in Article 90. Therefore, when reviewing the practical questions pertaining to Article 90, understand that these question are based on the overall purpose of the code, its format, and how it should be translated.

Chapter 1 provides general information for all electrical installations. Of all nine chapters, Chapter 1 consists of only two articles and is therefore the smallest. So let it be said now, once the contents of a chapter are understood, the next thing to do is read the scope of each article to understand the objective and subject matter of the article. For instance, when studying the practical questions relating to Article 100, you will be covering definitions that only provide means for interpreting the code. To receive instructions pertaining to an electrical installation, you must first learn to speak the language. Understanding the contents of a chapter and the scope of an article is very important. Information understood should lead to information comprehended.

Chapter 2 provides information relating to wiring and protection requirements. It includes articles that identify electrical conductors, branch and feeder circuits for both inside and outside installations, load calculations, services, overcurrent protection, and grounding and bonding. The articles containing these subject matters are those most-often used. Take time to realize how certain articles are linked together. For instance, Article 210 provides general requirements for branch circuits, while Article 215 provides general requirements for feeder circuits supplying branch circuit loads, and Article 230 covers conductors and equipment relating to service installation requirements. Now, take notice of how all three articles are linked to Article 220, which provides requirements for calculating branch-circuit, feeder, and service loads. Therefore, when reviewing the practical questions relating to the articles of Chapter 2, all practical questions will pertain to wiring and protection requirements.

Chapter 3 provides the information gathered from the articles of Chapter 2 wherein it supplies information relating to wiring methods and the materials commonly affiliated with electrical installations. Like Chapter 2, Chapter 3 is also used frequently because of its contents and articles that provide wiring methods and material applications. Of all nine chapters, Chapter 3 is the largest. It applies to wiring and installation methods for conductors and raceways—refer to Articles 300 and 310 specifically. Table 310.16 of Article 310 is probably the most referenced table of the entire NEC. For wiring enclosures such as metal and non-metal boxes and conduit bodies (Types LB, LL, C, etc.), Article 314 is the source. As it pertains to the various types of electrical cables and raceway that are used in performing the electrical installations, Articles 320–398 apply. Pay close attention and make sure you understand the differences between each type of material. Although armored cable (Type AC [BX]) and metal-clad cable (Type MC) appear to be the same in physical characteristics, they are as different as rigid metal conduit and rigid nonmetallic conduit, and electrical metallic tubing and electrical nonmetallic tubing. Study each individual article and related practical questions with the objective of realizing that each is different although they appear to be similar. Remember, if they weren't different, there wouldn't be different articles.

Chapter 4 covers equipment that is generally used, such as switches, receptacles, panelboards, appliances, motors, air-conditioning and refrigeration, generators, transformers, etc. Perhaps the most referenced article of Chapter 4 is Article 430, which covers electrical motors and supporting circuitry. Tables 430.248 and .250 are, without a doubt, two of the most referenced tables of Chapter 4. Articles 440 (Air-Conditioning and Refrigeration) and 450 (Transformers) are also frequently used.

Chapter 5 covers special occupancies. It is considered the most difficult chapter to understand, mainly because it includes hazardous (classified) locations, which are covered in Articles 500–510 and followed by a combination of articles covering certain occupancies that will include classified locations such as commercial garages, aircraft hangers, and even hospitals. Chapter 5 also includes occupancies such as theaters (Article 520), carnivals (Article 525), mobiles homes (Article 550), and marinas and boatyards (Article 555), to name a few. Performing an electrical installation for most occupancies in Chapter 5 or the application of a specific type wiring method may never occur for most electrical professionals, but such is quite common. Nevertheless, all articles covered under Chapter 5 are still a part of the code and must be studied with the same intensity as other code articles that are considered easier to understand. Even if you can't quite grasp the information provided in a practical question, always try to remember where the information was found.

Perhaps, like Chapter 5, the articles covered under Chapter 6 that pertains to special equipment may seldom be used or applied, yet as a whole they are more appealing, which results in greater acceptance. You might find that the practical questions relating to the articles of Chapter 6 in comparison to the articles of Chapter 5 may prove to be more easily received.

Chapter 7 covers special conditions where Articles 700–708 relate to emergency power systems and Articles 720–770 relate to low-voltage circuits, fire alarm systems, and optical fiber applications. Although these articles and the related practical questions may seem distant, just remember that upon completing these practical questions, your level of knowledge will have increased.

Chapter 8 specifically covers communications systems such as communication circuits, radio and television equipment, and distribution systems. The only drawback with these subject matters from an electrical professional's perspective is, where do they fit in? But once again, let me remind you that the practical questions relating to Articles 800–830 are just as important.

Finally, Chapter 9 and the annexes need your special attention to ensure that you are familiar with the information and data provided concerning raceway fills and conductor properties. Overall, for referencing and calculation purposes, these tables—along with the annexes—are good sources to retain.

Therefore, in concluding, let me again reemphasize the importance of understanding the contents of all chapters and the scope of each article. And now I can finally say, the practical questions are all yours. Have fun!

# CHAPTER 1–GENERAL

## ● ARTICLE 90–INTRODUCTION

1. By special permission, the authority having jurisdiction may _____ specific requirements in the code.

   a. amend          b. supersede          c. waive          d. extend

2. A direct mathematical conversion that involves a change in the description of an existing measurement but not in the actual dimension is considered a(n) _____.

   a. alternate conversion

   b. soft conversion

   c. temporary conversion

   d. hard conversion

3. The purpose of the National Electrical Code is _____.

   a. to provide instructions for those individuals desiring to become electricians

   b. to serve as a design manual for engineers and designers

   c. for the practical safeguarding of persons and property

   d. to assist legal and forensic experts in civil courts

4. Fine print notes are code enforceable.

   a. True          b. False

5. The National Electrical Code covers the installation of _____.

   a. railway rolling stock

   b. automotive vehicles

   c. fiber optic cables and raceways

   d. self-propelled mobile surface mining machinery

# ⦿ ARTICLE 100–DEFINITIONS

1.  A circuit breaker that has no time delay in its tripping acting is a(n) _____ breaker.

    a. nonadjustable      b. instantaneous trip      c. inverse time      d. adjustable

2.  Varying duty requires _____.

    I. operation in which the load conditions are at regular intervals

    II. operation in which the load conditions are at intervals of time

    III. operation in which the load conditions are at alternate intervals

    a. only I      b. only II      c. only III      d. only II and III      e. only I and III

3.  An enclosure that is sized to allow personnel to reach into for maintaining equipment is called a _____ enclosure.

    a. manhole      b. junction box      c. handhole      d. wiring

4.  A(n) _____ is a point on the wiring system at which current is taken to supply utilization equipment.

    a. utilization outlet      b. outlet      c. power outlet      d. receptacle outlet

5.  When service equipment is attached to a pole, it is considered a structure.

    a. True      b. False

6.  The definition of _____ is a connection used to establish electrical continuity and conductivity.

    a. grounding      b. bonding      c. splicing      d. pressure connections

7.  A system or circuit conductor that is intentionally grounded is called a(n) _____.

    a. grounding conductor      b. grounded conductor      c. ground
    d. equipment grounding conductor

8.  An assembly of one or more enclosed sections having a common power bus and principally containing motor control units is recognized in the National Electrical Code as a _____.

    a. load center      b. panelboard      c. motor control center      d. switchboard

9. The underground service conductors between the street main and the first point of the connection to the service-entrance conductors in a meter inside or outside a building wall are defined as _____.

    a. underground system      b. service loop      c. service lateral      d. permanent service

10. "Capable of being reached quickly for operation, renewal, or inspections without having to climb over, remove obstacles, or resort to portable ladders" is the definition of _____.

    a. accessible (wiring method)      b. accessible, readily      c. accessible (equipment)
    d. accessible

11. A load where the wave shape of the steady-state voltage does not follow the wave shape of the applied current is called a nonlinear load.

    a. True                b. False

12. A single unit providing complete and independent living facilities for one or more persons including permanent provisions for living, sleeping, cooking, and sanitation is considered a _____.

    a. one-family dwelling      b. dwelling unit      c. multifamily dwelling
    d. one or two-family dwelling unit

13. A hoistway is a _____ in which an elevator or dumbwaiter is designed to operate.

    a. vertical space or opening      b. shaftway      c. hatchway or well hole
    d. all choices apply

14. Any current in excess of the rated current of equipment or the ampacity of a conductor is called _____.

    a. overload      b. overcurrent      c. infinite      d. continuous

15. A _____ switch is one that is rated in amperes and is capable of interrupting its rated current at its rated voltage.

    a. three-way      b. general-use      c. general-use snap      d. disconnect

16. A kitchen is an area with a dishwasher and garbage disposal that provides permanent means for cooking.

    a. True                b. False

17. A _____ installation is something that is so constructed that moisture will not enter the enclosure under specified test conditions.

    a. moistureproof      b. weatherproof      c. watertight      d. sealproof

18. "When consisting of necessary equipment such as switch(es), circuit breaker(s), or fuse(s) that are connected to the load end of service conductors to a building, structure, or designated area where the intentions are to control and cut off the supply" best describes _____.

    a. overcurrent protection      b. service equipment      c. control devices
    d. service disconnecting means

19. A wet location involves installations that are subject to saturation with water or other liquids.

    a. True      b. False

20. "Not guarded by locked doors, elevation, or effective means in which close approach is admitted" is defined as _____.

    a. accessible (as applied to equipment)      b. accessible, readily      c. accessible
    d. accessible (as applied to wiring methods)

21. Even though they may become accessible by withdrawing, wires in concealed raceways are considered _____.

    a. inaccessible      b. protected      c. concealed      d. removable

22. When electrically connected to a source of current, a system is considered energized.

    a. True      b. False

23. The phrases "capable of being inadvertently touched" and "behind panels designed to allow access" describe the term _____.

    a. exposed (as applied to live parts)      b. exposed (as applied to wiring methods)
    c. either a or b      d. both a and b

24. A fault such as a short circuit or ground fault is considered an overload.

    a. True      b. False

25. A thermal protector may consist of one or more sensing elements integral with the motor or motor-compressor and an external control device.

    a. True      b. False

26. "Where intended for use in parallel with an electric utility to supply common loads that may deliver power to the utility" is the definition of a _____.

    a. converter      b. distribution system      c. network      d. utility-interactive inverter

27. A(n) _____ is a branch circuit that supplies two or more receptacle or outlets for lighting and appliances.

    a. appliance branch circuit             b. general purpose branch circuit
    c. individual branch circuit            d. isolated branch circuit

28. _____ can fulfill the requirements of weatherproofing where varying weather conditions are not a factor.

    a. Watertight          b. Raintight          c. Rainproof          d. All choices apply

29. _____ devices are provided with interrupting ratings appropriate for the intended use (but no less than 5,000 amperes).

    a. Switching       b. GFCI       c. Branch-circuit overcurrent       d. Circuit breaker

30. The current that a conductor can carry continuously without exceeding its temperature rating is best described by the term _____.

    a. amperes          b. short-circuit          c. ampacity          d. electricity

31. A conductor encased within material of composition or thickness that is not recognized by the National Electrical Code as electrical insulation is called a(n) _____ conductor.

    a. covered          b. insulated          c. wrapped          d. enclosed

32. The highest current at rated voltage that a device is intended to interrupt under standard test conditions is called the _____.

    a. short-circuit capacity       c. Underwriters Lab test results       b. interrupting rating
    d. root-mean-square

33. A circuit's voltage is the _____ root-mean-square (rms) difference of potential between any two conductors of the circuit concerned.

    a. least          b. highest          c. greatest          d. maximum

34. A person who has received safety training, is aware of surrounding hazards, and has skills and related knowledge pertaining to the construction and operation of electrical equipment and installations is considered _____.

    a. competent          b. skilled          c. qualified          d. certified

35. When the words "Thermally Protected" appear on a nameplate of a motor-compressor or motor, this is an indication that the motor is provided with a(n) _____.

    a. overcurrent device    b. short-circuit rating    c. thermal protector    d. separate winding

36. A lampholder itself is not a luminaire.
   a. True                b. False

37. Intermittent duty requires an alternate interval of _____.
   a. load and rest      b. load, no load, and rest      c. load and no load      d. all choices

38. For a circuit to be regarded as a multiwire branch circuit, it must have _____.
   I. a grounded conductor that has equal voltage between it and each ungrounded conductor of the circuit
   II. two or more ungrounded conductors that have a voltage between them
   III. the grounded conductor connected to the neutral or grounded conductor of the system
   a. only I      b. only II      c. only III      d. neither I, II, nor III      e. I, II, and III

39. "Where no adjustment or altering of the value of current at which a breaker will trip" is the definition of a(n) _____ breaker.
   a. inverse time      b. nonautomatic      c. one-time      d. nonadjustable

40. A plug is a device that, by inserting in a receptacle, establishes a connection between the conductors of the attached flexible cord and the conductors connected permanently to the receptacle.
   a. True                b. False

41. When the maximum current required by a load is expected to continue for _____ hours or more, it is recognized as a continuous load.
   a. 10      b. 7      c. 5      d. 3

42. A _____ consists of all circuit conductors between the service equipment, the source of a separately derived system or other power supply source, and the final branch–circuit overcurrent device.
   a. branch circuit      b. panelboard      c. feeder      d. motor control center

43. "Equipment or materials included in a list published by an organization that is acceptable to the authority having jurisdiction" is the definition for the term _____.
   a. identified      b. approved      c. labeled      d. listed

44. "Where treated, protected, or constructed to prevent rain from interfering with the successful operation of the apparatus under specified test conditions" is the definition of _____.
   a. sealtight      b. rainproof      c. watertight      d. raintight

45. Utilization equipment uses electric energy for _____ compartments.

    a. electronic     b. lighting     c. heating     d. all choices apply

46. At the _____ of the system, the vectorial sum of the nominal voltages from all other phases within the system that utilize the neutral, with respect to the neutral point, is zero potential.

    a. grounded conductor     b. service entrance     c. neutral point     d. supply side

47. The overhead service conductors from the last pole or other aerial support to and including splices, if any, that connect the service-entrance conductors to buildings or other structures are called the _____.

    a. point of attachment     b. service drop     c. overhead service     d. service-entrance

48. Operation at a substantially constant load for an indefinitely long time is called _____.

    a. indefinite     b. continuous load     c. non-varying     d. continuous duty

49. An enclosure that is designed for either flush or surface mounting that's provided with a trim, mat, or frame on which a swinging door or doors can be hung is called a(n) _____.

    a. container     b. panelboard     c. box     d. cabinet

50. For an area to be considered a bathroom, it must include _____.

    a. a tub or shower
    b. a toilet and shower
    c. a shower and basin
    d. a tub and toilet

51. The ratio of the maximum demand of a system, or part of a system, to the total connected load of a system or the part of the system under consideration is called a demand factor.

    a. True     b. False

52. Compression and set-screw connectors are considered _____.

    a. devices     b. fittings     c. equipment     d. components

53. If partially protected under a canopy, such place is considered a _____ location.

    a. moist     b. damp     c. dry     d. sultry

54. "Where protected or constructed so that exposure to a beating rain will not result in the entrance of water under specified test conditions" best describes the term _____.

    a. weathertight        b. rainproof            c. floodproof           d. raintight

55. "Equipment or materials to which has been attached an identifying mark of an organization that is acceptable to the authority having jurisdiction" is considered _____.

    a. approved            b. labeled              c. listed               d. identified

56. A value assigned to a circuit or system for the purpose of conveniently designating its voltage class is called the _____ voltage.

    a. operating           b. nominal              c. actual               d. utility

57. Constructed so that dust will not interfere with successful operation is called dusttight.

    a. True                b. False

58. A _____ is used to connect the system grounded conductor or the equipment to a grounding electrode or to a point on the grounding electrode system.

    a. ground rod          c. grounding clamp              b. grounding electrode conductor
    d. ground ring

59. "Where storage facilities within a compartment are provided, along with accommodations for sanitary, sleeping, and living" identifies a guest suite.

    a. True                b. False

60. The connection between the grounded circuit conductor and the equipment grounding conductor is a(n) _____.

    a. bonding jumper

    b. bonding conductor

    c. equipment bonding jumper

    d. main bonding jumper

61. An example of a device is a _____.

    a. pull box            b. conduit body         c. switch               d. locknut

62. A ground is a connection to earth or to some conducting body that serves in place of the earth.

    a. True                b. False

63. Where temporarily subject to dampness or wetness, such place is considered a _____ location.

    a. damp               b. dry               c. wet               d. moist

64. The conductors and equipment for delivering electric energy from the serving utility to the wiring system of the premises are recognized by the code as a service.

    a. True               b. False

65. Class A ground-fault circuit interrupters trip when the current to ground has a value in the range of 4 mA to 6 mA.

    a. True               b. False

# ● ARTICLE 110–REQUIREMENTS FOR ELECTRICAL INSTALLATIONS

1. To acquire protection that is substantially equivalent to the wall(s) of equipment, unused openings in _____ shall be effectively closed.

   a. raceways      b. cabinets      c. gutters      d. boxes
   e. all choices are applicable

2. What is the minimum working clearance for a circuit 277 volts to ground where exposed live parts exist on one side of the working space and grounded parts on the other side of the working space?

   a. 4'      b. 3.5'      c. 5'      d. 3'

3. Circuits for lighting and power shall not be connected to any system that contains trolley wires with a ground return.

   a. True      b. False

4. Only conductors rated for 60°C can be used where the termination provisions of equipment for circuits rated 100 amperes or less are unknown.

   a. True      b. False

5. A fence that is 7 feet in height is considered adequate for deterring access to enclosures exceeding 600 volts.

   a. True      b. False

6. Manhole covers are required to be over 45 lbs or otherwise designed to require the use of tools for opening.

   a. True      b. False

7. The work space about electrical equipment rated for 600 volts or more must permit at least a _____ opening of doors or hinged panels.

   a. 45°      b. 90°      c. 150°      d. 180°

8. The term watertight is mostly used in connection with Enclosure Types 4, 4X, 6, and 6X.

   a. True      b. False

9. Where rooms or other guarded locations contain exposed live parts, _____ must be made visible to alert unqualified persons from entering.

   a. emergency lights      b. warning signs      c. guards      d. monitors

10. Equipment and conductors required or permitted by the National Electrical Code shall be acceptable only if recognized by the authority having jurisdiction.

    a. True          b. False

11. Round excess openings to manholes must not be less than _____ inches in diameter.

    a. 26.5          b. 22 x 22          c. 650          d. no correct choice

12. For an existing one-family dwelling, the minimum headroom space is allowed to be less than 6.5 feet where service equipment or panelboards do not exceed _____.

    a. 150A          b. 200A          c. 100A          d. 400A

13. A manhole containing optical fiber cables is permitted to have one of the horizontal work space dimensions reduced to _____, where the other horizontal clear work space is increased so the sum of the two dimensions is not less than _____.

    a. 6", 2'          b. 2', 6'          c. 6', 2"          d. 6', 2'

14. The material and sizes given in the code shall apply to _____ conductors when the conductor material is not specified.

    a. aluminum and copper          b. copper-clad aluminum          c. copper          d. aluminum

15. Where associated with potentially energized circuits, unused current transformers must be _____.

    a. removed          b. disconnected          c. short-circuited          d. de-energized

16. The minimum distance from a fence to a distribution transformer rated for 24,000 volts is _____.

    a. 10'          b. 15'          c. 20'          d. 25'

17. A switchboard that is 6' in width and rated for 4160 volts requires a minimum of _____ entrance(s).

    a. 2          b. 3          c. 1          d. 4

18. Faults are assumed to occur between two or more circuit conductors or between a circuit conductor and the grounding conductor or enclosing metal raceway.

    a. True          b. False

19. The voltage rating of electrical equipment shall not be less than the nominal voltage of a circuit to which it is connected.

    a. True          b. False

20. The conductor having the higher phase voltage to ground on a 4-wire, delta-connected system must be permanently identified with an outer finish that is _____ in color.

    a. blue         b. red         c. orange         d. black

21. Unless a device is identified for the purpose of intermixing conductors made of dissimilar metals, it cannot be used.

    a. True         b. False

22. Where driven into holes in masonry, concrete, or plaster, _____ shall not be used to secure electrical equipment to the surface to which it is mounted.

    a. nails         b. anchor bolts         c. wooden plugs         d. plastic brackets

23. A dropped ceiling that does not add strength to a building's structure must not be considered a structural ceiling.

    a. True         b. False

24. Some cleaning and lubricating compounds can cause severe _____ of many plastic materials used for insulating and structural applications in equipment.

    a. damage         b. defects         c. deterioration         d. cracking

25. Conductor sizes are expressed in either American Wire Gage (AWG) or circular mils.

    a. True         b. False

26. High-voltage conductors in tunnels shall be installed in metal conduit or other metal raceway.

    a. True         b. False

27. Equipment intended to interrupt current at _____ shall have an interrupting rating sufficient for nominal circuit voltage and current that is available at the line terminals of the equipment.

    a. disconnecting means    b. other than fault levels    c. overcurrent device    d. fault levels

28. A type 3RX enclosure is required where the following environmental condition occurs:

    a. failing dirt         b. snow         c. corrosive agents         d. prolonged submersion

29. Consideration of _____ must be evaluated when judging equipment.

   a. electrical insulation

   b. arcing effects

   c. mechanical strength and durability

   d. wire-bending and connection space

   e. all of the preceding

   f. none of the preceding

30. _____ ladders used to gain access to a vault or tunnel shall be corrosion resistance.

   a. Step          b. Adjustable          c. Fixed          d. Steel

31. Connections by means of wire-binding screws or studs and nuts that have upturned lugs shall be permitted for _____ or smaller conductors.

   a. No. 12        b. No. 6          c. No. 10          d. No. 8

32. Placing electrical equipment on elevated platforms to exclude unqualified persons is a recognized means for guarding against accidental contact.

   a. True          b. False

33. When completed, all wiring installations shall be free from _____.

   a. ground fault   b. short circuits   c. either a or b   d. both a and b

   e. neither a nor b

34. The minimum elevation for unguarded live parts rated for 125kV above working space is approximately _____.

   a. 9.5'          b. 12.25'          c. 13.5'          d. 14.75'

# CHAPTER 2–WIRING AND PROTECTION

## ● ARTICLE 200–USE AND IDENTIFICATION OF GROUNDED CONDUCTORS

1.  The identification of terminals to which a grounded conductor is to be connected shall be substantially _____ in color.

    a. gray                  b. white                  c. black                  d. silver

2.  An insulated grounded conductor of _____ or smaller must be identified by a continuous white or gray outer finish or by three continuous white stripes on other than green insulation along its entire length.

    a. No. 4 AWG          b. No. 6 AWG          c. No. 8 AWG          d. No. 2 AWG

3.  A white or gray conductor, where properly re-identified, can be used as a return conductor from a switch to a switched outlet.

    a. True                  b. False

4.  If not visible, terminals intended for connection to grounded conductors shall be _____.

    a. marked with the letter W

    b. marked with the word white

    c. colored white

    d. only a or b

    e. only b or c

    f. a, b, or c

5.  Insulated grounded conductors larger than No. 6 can be marked with a distinctive white marking that encircles the conductor or insulation where terminated.

    a. True                  b. False

6.  For shell screw devices, the terminal for grounded conductors shall be the one connected to the _____.

    a. silver terminal       b. shell screw       c. white conductor       d. all choices applicable

7.  As described, the term "electrically connected" means to be capable of carrying current by means of a physical connection opposed to current produced by induction.

    a. True                 b. False

8.  Multiconductor flat cable exceeding No. 6 AWG in size is permitted to have a(n) _____ on the grounded conductor.

    a. distinctive marking     b. external ridge     c. smooth edge     d. choices not applicable

# ⦿ ARTICLE 210–BRANCH CIRCUITS

1. When a 20-ampere branch circuit supplies more than one bathroom in a dwelling, other equipment can also be supplied from the circuit.

   a. True                b. False

2. A receptacle in a dwelling unit is required to be installed so that no point measured horizontally along the floor line in any wall space is more than _____ from a receptacle outlet.

   a. 3'                b. 6'                c. 9'                d. 12'

3. A 125 volt, 1φ, 15 or 20-ampere receptacle is required to be installed on the same level and within 75 feet of where HACR equipment is installed.

   a. True                b. False

4. Where the assembly, including the overcurrent devices protecting the branch circuit(s), is listed for operation at _____ percent of its rating, the ampere rating of the overcurrent device shall be permitted to be not less than the sum of the continuous load plus the noncontinuous load.

   a. 80                b. 100                c. 150                d. 175

5. In kitchens and dining rooms of dwelling units, receptacle outlets shall be installed at each wall counter space that is _____ or wider.

   a. 24"                b. 18"                c. 12"                d. 36"

6. If located within 18" of a wall, floor receptacles are required to be counted as a part of the required number of receptacle outlets.

   a. True                b. False

7. A heavy-duty lampholder shall have a rating of not less than 660 watts if of the admedium type and not less than 750 watts if of any other type.

   a. True                b. False

8. The rating of a 15A circuit that serves utilization equipment that is cord and plug connected must not exceed 50 percent of the circuit's ampere rating.

   a. True                b. False

9.  At least one receptacle outlet shall be installed directly above a show window for each _____ or major fraction thereof.

    a. 6 linear feet          b. 9 linear feet          c. 12 linear feet          d. 15 linear feet

10. A No. 10 conductor is used to supply a branch circuit. The conductor is protected by an overcurrent device rated for 60 amperes. What is the rating of the branch circuit?

    a. 25A          b. 30A          c. 35A          d. 60A

11. At least one receptacle outlet that is accessible while standing at grade level and not more than 6.5' above grade shall be installed at the _____ of a one-dwelling.

    a. rear          b. front          c. side          d. a and c          e. b and c          f. a and b

12. In dwelling units, a receptacle is required for each wall space that is _____ or more in width and unbroken along the floor line by doorways, fireplaces, and similar openings.

    a. 1'          b. 2'          c. 3'          d. 4'

13. Switched receptacles in other than kitchens and bathrooms can be used instead of lighting outlets in habitable rooms of dwellings.

    a. True          b. False

14. Multiwire branch circuits are not allowed to supply line-to-line loads.

    a. True          b. False

15. A receptacle that is installed in the garage of a dwelling unit is not required to be ground fault protected if not readily accessible.

    a. True          b. False

16. At least one wall receptacle outlet in bathrooms of dwelling units must be installed within _____ of the outside edge of each basin.

    a. 1'          b. 2'          c. 3'          d. 4'

17. A circuit rated for 40A can supply a receptacle rated for _____.

    a. 30A          b. 40A          c. 50A          d. 30A or 40A          e. 40A or 50A

18. All _____ receptacles installed on the rooftop of a commercial building are required to be ground-fault protected.

    a. 15A 120 volt          b. 15 and 20A, 125 volt          c. 15/20A, 250 volt          d. only b and c
    e. only a and b          f. only b

19. Other than the outlets permitted in NEC 210.52(B)(1), no other outlets are permitted to be on a small appliance branch circuit.

    a. True                 b. False

20. All multiwire branch-circuit conductors are required to originate from the same panelboard.

    a. True                 b. False

21. Before applying any correction factors or adjustment, the allowable ampacity of a branch circuit must not be less than the noncontinuous load(s) plus _____ percent of the continuous load(s).

    a. 80           b. 100              c. 125              d. 150

22. The auxiliary equipment of a metal halide lamp can be supplied by either 120 volts or 277 volts.

    a. True                 b. False

23. When a hallway in a dwelling unit is 6 feet or more in length, at least one receptacle must be installed.

    a. True                 b. False

24. The minimum size tap conductor that can be used on circuit protected by a 40A overcurrent device is a _____.

    a. No. 14           b. No. 12           c. No. 10           d. all of the preceding

25. In guest rooms of motels, the voltage shall not exceed 120 volts between conductors that supply the terminals of _____.

    a. cord and plug loads not exceeding 1440 volt-amperes

    b. cord and plug loads less than 1/4 horsepower

    c. lighting fixtures

    d. only a and b

    e. a, b, and c

    f. none of the preceding

26. For a cord and plug connected load, the maximum allowable load on a 30A circuit is _____.

    a. 24A           b. 30A              c. 35A              d. not applicable

27. All branch circuits that supply 125 volt, single phase, 15 and 20 ampere _____ in the bedroom of a dwelling must be protected by an arc fault circuit interrupter.

a. appliances        b. receptacles        c. outlets        d. cords

28. Receptacle outlets located in _____ must be supplied on a small appliance branch circuit.

a. breakfast rooms        b. pantries        c. dining rooms        d. kitchens
e. all choices apply

29. The use of an autotransformer is permitted without the connection to a grounded conductor when transforming from _____ to _____.

a. 120V, 240V        b. 208V, 240V        c. 240V, 480V        d. 480V, 600V

30. Which of the following branch circuits is considered a multiwire branch circuit?

a. a 3W, 3φ, 240 volt circuit    b. two 2W, 120 volt circuits    c. three 2W, 277 volt circuits
d. a 4W, 3φ, 208/120 volt circuit

31. For ranges rated 8.75kW or larger, the minimum branch-circuit rating shall be _____.

a. 30A        b. 35A        c. 40A        d. 45A

# ● ARTICLE 215–FEEDERS

1. Grounding means must be included or provided in accordance with NEC _____ when a feeder supplies branch circuits requiring equipment grounding conductors.

   a. 250.118          b. 250.122          c. 250.134          d. 250.142

2. When a feeder conductor carries the total load supplied by a service-entrance conductor with an ampacity of 50 amperes or less, the ampacity of the feeder conductor must not be less than that of the service-entrance conductor.

   a. True          b. False

3. The use of a common neutral with a feeder is allowed to supply _____.

   a. two sets of 3W feeders          b. two sets of 4W feeders          c. two sets of 5W feeders
   d. three sets of 3W feeders        e. choices not applicable           f. choices a–d

4. Where rated for a _____ or more and installed on a solidly grounded wye electrical system of more than 150 volts to ground but not exceeding 600 volts phase-to-phase, a feeder disconnect must be provided with ground fault protection.

   a. 500A          b. 750A          c. 1000A          d. 1200A

5. The maximum voltage drop recommended on both feeders and branch circuits combined to the farthest outlet should not exceed _____ percent.

   a. 3          b. 5          c. 7          d. 10

6. Before applying any correction factors or adjustment, the allowable ampacity of a feeder must not be less than the noncontinuous load(s) plus _____ percent of the continuous load(s).

   a. 115          b. 125          c. 150          d. 175

7. The rating of an overcurrent device supplying a feeder load should not be less the continuous load plus _____ of the continuous load.

   a. 80          b. 100          c. 125          d. 150

# ● ARTICLE 220–BRANCH-CIRCUIT, FEEDER, AND SERVICE CALCULATIONS

1.  In banks or office buildings, the receptacle loads shall be calculated at _____ voltamperes per square foot.

    a. 1            b. 2            c. 3            d. 3.5

2.  To use the derived results based on the optional calculation for a dwelling unit, a service conductor must have an ampacity of 100 amperes or more.

    a. True         b. False

3.  A 75 percent demand factor can be used when _____ or more appliances are fastened in place in dwelling units.

    a. 2            b. 3            c. 4            d. 5

4.  When computing the total farm loads, a demand factor of 65 percent must be used for calculating the _____ load.

    a. largest      b. second largest      c. third largest      d. fourth largest

5.  A service or feeder neutral load is permitted to be reduced to 70 percent for _____.

    a. household ranges    b. household dryers    c. a grounded conductor in excess of 200A
    d. only a and b        e. only c              f. all choices apply

6.  The computed load for circuits supplying lighting units with ballast shall be based on the total ampere ratings of the unit and not on the total wattage of the lamps.

    a. True         b. False

7.  Household cooking appliances rated _____ can use the demand factors listed in Table 220.55.

    a. less than 1.75kW    b. for 1.75kW    c. over 1.75kW    d. choices not applicable

8.  The demand load for a dishwasher having a wash and drying cycle only requires the cycle that produces the largest load to be considered for computing the overall load.

    a. True         b. False

9.  One allowance for determining the actual demand of an existing installation for calculating a feeder load is to use data of the maximum demand covering a _____ period.

    a. 3-month      b. 6-month      c. 9-month      d. 12-month

10. When a computation results in a fraction of an ampere that is less than _____, such fraction is permitted to be dropped.

    a. .3                b. .5                c. .7                d. .8

11. Table 220.55 must not be used for household cooking appliances rated over 1.75kW when used in schools for teaching home economics.

    a. True              b. False

12. The demand factor for 35-unit apartment complex using the optional calculation is _____.

    a. 24 percent        b. 30 percent        c. 38 percent        d. 45 percent

13. A minimum of _____ volt-amperes shall be computed for each sign outlet branch circuit.

    a. 1000              b. 1200              c. 1500              d. 2000

14. The minimum load for household electric dryers must be based on _____.

    a. the dryer nameplate rating

    b. 5000W

    c. the line to line voltage

    d. 5000 watts or the nameplate rating, whichever is larger

15. In determining the number of branch circuits for general illumination, the demand factors of Table 220.42 must be applied.

    a. True              b. False

16. Under certain conditions, the authority having jurisdiction can grant permission for feeder and service conductors to have an ampacity less than 100 percent for fixed electric space heating loads.

    a. True              b. False

17. Where eight commercial kitchen appliances are used, the overall demand load can be reduced to _____ percent.

    a. 25                b. 40                c. 65                d. 80

18. Where track lighting is used in guest rooms of hotels, 150VA shall be included for every _____ of lighting track.

    a. 2 amps            b. 2 ft²             c. 2 ft              d. 2 watts

19. For circuits supplying loads consisting of motor-operated utilization equipment that is fastened in place and has a motor larger than _____ in combination with other loads, the total computed load shall be based on 125 percent of the largest motor load plus the sum of the of the other loads.

   a. 1/8 hp             b. 1/4 hp             c. 1/2 hp             d. 3/4 hp

20. In dwelling units, a garage must be included in the computed floor area.

   a. True               b. False

21. Small appliance and laundry loads in apartments must be computed at 1500VA for each _____ circuit small appliance branch circuit.

   a. splitwire          b. three-wire         c. multiwire          d. two-wire

22. Once a receptacle load exceeds 10kVA in a nondwelling unit, the remaining portion of the receptacle loads can be reduced to _____ percent.

   a. 25                 b. 50                 c. 75                 d. 80

23. Receptacle outlets shall be computed at not less than _____ volt-amperes for each single or for each multiple receptacle on one yoke.

   a. 90                 b. 180                c. 200                d. 600

# ● ARTICLE 225–OUTSIDE BRANCH CIRCUITS AND FEEDERS

1. When conductors rated for 2.4kV are installed over balconies, the minimum vertical clearance must be _____.
   a. 10'        b. 13.5'        c. 18.5'        d. 20'

2. In a multiple-occupancy building, each occupant shall have access to the occupant's supply disconnecting means.
   a. True        b. False

3. Trees are permitted to be used for supporting spans of overhead conductors.
   a. True        b. False

4. A set of 240 volt open conductors are installed over the parking lot of a shopping mall. The conductors must have clearance of _____.
   a. 10'        b. 12'        c. 15'        d. 18'

5. Additional feeders or branch circuits are permitted when the capacity requirements are in excess of _____ at a supply voltage of 600 volts or less.
   a. 1000A        b. 1500A        c. 2000A        d. 3000A

6. In a commercial building, a feeder or branch circuit disconnecting means must have a rating of not less than _____.
   a. 60A        b. 100A        c. 150A        d. 200A

7. Open conductors must be separated from open conductors of other circuits or systems by not less than _____.
   a. 2"        b. 4"        c. 6"        d. 12"

8. Conductors that are installed above the top level of a window are permitted to be less than the _____ requirement.
   a. 12"        b. 24"        c. 36"        d. 48"

9. Overhead conductors for festoon lighting shall not be smaller than _____ unless the conductors are supported by messenger wires.
   a. No. 14        b. No. 12        c. No. 10        d. No. 8

10. Raceways on the exterior surfaces of buildings or other structures shall be _____ and arranged to _____.

   a. secure, prevent damage

   b. raintight, drain

   c. weatherproof, eliminate moisture

   d. sealtight, prevent condensation

11. Open individual overhead conductors must be insulated or covered when within 10 feet of any building or structure other than supporting poles or towers.

   a. True                    b. False

# ● ARTICLE 230–SERVICES

1. According to the NEC, service-lateral conductors are permitted to be spliced or tapped.

   a. True          b. False

2. Each service disconnect shall simultaneously disconnect all ungrounded service conductors that it controls from the premises wiring system.

   a. True          b. False

3. Where the voltage exceeds _____ between conductors that enter a building, they shall terminate in a metal-enclosed switchgear compartment or vault.

   a. 600 volts      b. 13,000 volts      c. 24,000 volts      d. 35,000 volts

4. Fuel cell systems are permitted to be connected on the supply side of service disconnecting means.

   a. True          b. False

5. Open service conductors not exposed to weather and rated less than 600 volts must maintain a _____ clearance between conductors as a minimum.

   a. 12"      b. 6"      c. 3"      d. 2.5"

6. The minimum size overhead service conductor allowed by the NEC is a _____.

   I. No. 6 copper AWG

   II. No. 8 aluminum

   III. No. 6 copper-clad or aluminum

   IV. No. 8 copper AWG

   a. I only      b. II only      c. III only      d. IV only      e. I and II
   f. II and III      g. III and IV      h. I and IV

7. To prevent the entrance of moisture, drip loops must be formed on each individual service entrance conductor.

   a. True          b. False

8. The maximum delay for ground-fault currents equal to or greater than 3000 amperes shall be _____.

   a. 60 seconds      b. 30 seconds      c. 5 seconds      d. 1 second

9. The maximum number of disconnects for a set of service entrance conductors is _____.
   a. 2                  b. 7                  c. 5                  d. 6

10. Unless permitted, a building or other structure served shall be supplied by only one service.
    a. True              b. False

11. The minimum size service disconnecting means for a single family house is _____.
    a. 60A               b. 80A               c. 100A              d. 125A

12. A terminal or bus to which all grounded conductors can be attached by means of pressure connectors are permitted when the service disconnecting means does not disconnect the grounded conductor from the premises wiring.
    a. True              b. False

13. Where vacuum circuit breakers constitute the service disconnecting means, a(n) _____ switch which visible break contacts shall be installed on the supply side of the disconnecting means and all associated service equipment.
    a. fused             b. double pole       c. isolating         d. dedicated

14. Service disconnecting means are not permitted to be installed in _____.
    a. storage rooms     b. bedrooms          c. bathrooms         d. garages

15. Where used for overhead service, the grounded conductor of a multiconductor cable is permitted to be _____.
    a. insulated         b. white             c. bare              d. gray

16. No overcurrent device shall be inserted in a grounded service conductor except a _____ that simultaneously opens all conductors of the circuit.
    a. fuse              b. contactor         c. circuit breaker   d. control device

17. When voltage between conductors does not exceed _____ and the slope of a roof is 4":12" or greater, the vertical clearance of service drop conductors can be reduced 3 feet.
    a. 120V              b. 208V              c. 240V              d. 300V

18. Where subject to physical damage, service cable must be protected by _____.
    a. Schedule 80, PVC          b. intermediate metal conduit      c. electrical metallic tubing
    d. rigid metal conduit       e. all of the preceding            f. only b and d

19. In no case shall the point of attachment of service conductors to a building be less than _____ above finished grade.

    a. 8'              b. 9'              c. 10'              d. 12'

20. Service conductors supplying a building or structure are allowed to pass through the interior or another building or structure.

    a. True              b. False

21. When installed in an auxiliary gutter, the grounded conductor of a set of service entrance conductors is permitted to be bare.

    a. True              b. False

22. A separate set of service entrance conductors are permitted to be installed in a multifamily complex to supply lighting and fire alarm branch circuits.

    a. True              b. False

23. The _____ conductor for the solidly grounded wye system shall be connected directly to ground through a grounding electrode system, as specified in 250.50, without inserting any resistors or impedance device.

    a. grounding        b. grounded        c. bare grounded        d. insulated grounded

24. Service conductors installed as open conductors must have a clearance of not less than _____ from porches and fire escapes.

    a. 1'              b. 3'              c. 5'              d. 7'

25. Protection for ungrounded service conductors must be provided by an overcurrent device that is in series with each ungrounded conductor.

    a. True              b. False

26. A raintight service head must be provided for _____.

    a. service cable        b. service raceway        c. both a and b        d. choices not applicable

27. Service conductors are required to be attached to the service disconnecting means by _____ or other approved means.

    a. clamps              b. pressure connectors        c. solder        d. only a and b
    e. only b and c        f. only a and c

28. A set of 480 volt service conductors are installed between two buildings over an alley. The conductors must have vertical clearance of _____.

    a. 13'                    b. 15'                    c. 18'                    d. 21'

29. Before applying any correction factors or adjustment, the allowable ampacity of a service-entrance conductor must not be less than the noncontinuous load(s) plus _____ percent of the continuous load(s).

    a. 115                    b. 125                    c. 150                    d. 200

30. Only _____ are allowed to the be installed in the same raceway or service cable with service conductors.

    I. load management control conductors

    II. grounding conductors

    III. bonding jumpers

    a. only II                b. only III               c. only I                    d. only II and III
    e. I, II, and III are applicable              f. I, II, and III are not applicable

# ● ARTICLE 240–OVERCURRENT PROTECTION

1. Supplementary overcurrent devices are required to be readily accessible.

   a. True            b. False

2. No overcurrent device shall be connected in series with any conductor that is intentionally grounded, unless _____.

   a. it will open both ungrounded and grounded conductors at the same time

   b. it is in a supervised location

   c. rated for 100A or less

   d. supplemented with ground fault protection

3. An Edison-base plug fuse is not classified for over _____ amperes.

   a. 15       b. 20       c. 25       d. 30

4. A combination of a current transformer and overcurrent relay shall be considered equivalent to a(n) _____.

   a. overcurrent device     b. circuit breaker     c. fuse     d. overcurrent trip unit

5. Feeder conductors are only allowed to be tapped where a circuit breaker or fuse exist.

   a. True            b. False

6. Plug fuses can be used with a 240/120V, 3φ, 4W closed delta system.

   a. True            b. False

7. A set of 17' tap conductors are supplied by No. 4/0 THWN copper conductors. The feeder is protected by a 250A overcurrent device. The minimum ampacity of the tap conductors must be _____.

   a. 65A       b. 84A       c. 168A       d. 230A

8. On a 2400V, 3φ system circuit breakers used for overcurrent protection of 3-phase circuits shall have a minimum of _____ overcurrent relay elements operated from three current transformers.

   a. one       b. two       c. three       d. six

9. The screw shell of a plug-type fuseholder shall be connected to the _____ side of the circuit.

a. supply          b. grounded          c. load          d. ungrounded

10. Overcurrent protection is provided for conductors and equipment to open the circuit if the current reaches a value that will cause an excessive or dangerous temperature in conductors or conductor insulation.

a. True          b. False

11. A No. 4/0 THWN copper conductor is used to supply a 145 load. What size overcurrent protection device is permitted per NEC 240.4(B)?

a. 150A          b. 175A          c. 200A          d. 250A

12. A similar change is permitted to be made in the size of an ungrounded conductor when a change occurs in the size of a grounded conductor.

a. True          b. False

13. Fuseholders of the Edison-base type shall be installed only where they are made to accept Type _____ fuses by the use of adapters.

a. A          b. F          c. S          d. W

14. Cartridge fuses and fuseholders shall be classified according to voltage and amperage ranges.

a. True          b. False

15. Unless specifically permitted, a No.10 aluminum conductor must be protected by a _____ overcurrent device.

a. 15A          b. 20A          c. 25A          d. 30A

16. _____ are recognized by the NEC as approved means for protecting overcurrent devices from physical damage.

a. Cutout boxes          b. Cabinets          c. both a and b          d. neither a nor b

17. A No.10 fixture wire is permitted to be tapped to a branch circuit conductor that is protected by a _____ overcurrent device.

a. 30A     b. 40A     c. 50A     d. a or b     e. b or c     f. a, b, or c

18. All fuses, fuseholders, and adapters shall be marked with its _____ rating.

a. voltage          b. short-circuit          c. ampere          d. AC/DC

19. A circuit breaker marked for 120/240V is permitted to be used on a solidly grounded circuit when the operating voltage of any conductor to ground does not exceed the lower voltage as marked.

    a. True                 b. False

20. In accordance with Article 240, a conductor other than a service conductor that has overcurrent protection ahead of its point of supply that exceeds the value permitted for similar conductors that are protected is called a _____ conductor.

    a. feeder           b. parallel                    c. tap                    d. secondary

21. A circuit breaker must be capable of opening all grounded and ungrounded conductors.

    a. True                 b. False

22. Unless otherwise permitted, overcurrent devices are required to be _____.

    a. accessible only to qualified personnel       b. accessible          c. readily accessible
    d. located in a secure environment

23. Type S fuses shall not be classified for over _____ volts.

    a. 110                 b. 115                    c. 125                    d. 130

24. In an area of a single family house where only a toilet and bathtub exist, overcurrent devices are permitted.

    a. True                 b. False

25. An 8' tap conductor should not extend beyond the equipment it supplies.

    a. True                 b. False

26. A circuit breaker having an interrupting rating other than _____ amperes shall have its interrupting rating shown on the circuit breaker.

    a. 500                 b. 2500                    c. 5000                    d. 10,000

27. A 75kVA 3ϕ transformer is rated 480-208/120V. Are the transformer's secondary conductors allowed to be protected by the primary overcurrent device?

    a. Yes                 b. No

28. When a feeder is provided in a building where automobiles are made and the building has a ceiling that is 51 feet in height, the total length of tap conductors, where used, must not exceed _____ feet.

    a. 35                 b. 60                    c. 75                    d. 100

29. Circuit breakers and fuses are permitted to be connected in parallel when _____.
    a. they are factory assembled in parallel      b. they protect motor control circuits
    c. they are listed as a unit      d. a and b      e. a and c      f. a, b, and c

30. The next higher standard overcurrent device above the ampacity of the conductors being protected is permitted as long as the rating of the overcurrent device does not exceed _____ amperes.
    a. 450      b. 601      c. 800      d. 1000

31. Individual single-pole circuit breakers, with or without approved handle ties, are permitted as protection for each ungrounded conductor of _____ branch circuits that serve only single-phase, line-to-neutral loads.
    a. individual      b. multiwire      c. appliance      d. general-purpose

32. In hotels and other similar locations, overcurrent devices are permitted to be located in a clothes closet.
    a. True      b. False

33. Cartridge fuses shall be marked to show _____.
    a. voltage rating
    b. current limiting capabilities
    c. manufacturer's trademark or name
    d. ampere rating
    e. interrupting rating where other than 10000A
    f. all choices apply

34. Under most conditions, overcurrent protection must be located at the point where conductors receive their supply and must be provided in each ungrounded circuit conductor.
    a. True      b. False

35. Circuit breakers shall clearly indicate whether they are in the _____ or _____ position.
    a. on "closed"      b. open "off"      c. closed "on"      d. off "open"
    e. a and b      f. b and c      g. c and d      h. a and d

36. Enclosures for overcurrent devices shall be mounted in a _____ position.
    a. horizontal      b. vertical      c. both choices applicable      d. neither choice applicable

37. For an industrial installation, all overcurrent devices are required to be _____ when the length of a transformer's secondary conductors are less than 25 feet in length.

   a. fuses            b. grouped            c. circuit breakers            d. within sight

38. Which fuse is not considered a standard ampere rating?

   a. 1A            b. 5A            c. 10A            d. 15A

# ● ARTICLE 250–GROUNDING AND BONDING

1. The grounded conductor shall not be smaller than the required _____.

    a. phase conductors     c. equipment grounding conductor     b. grounding electrode
    d. grounding electrode conductor

2. A metal underground water pipe permitted as a grounding electrode must be in direct contact with the earth for _____ or more and electrically continuous to the points of connection of the grounding electrode conductor and the bonding conductors.

    a. 5 feet        b. 8 feet        c. 10 feet        d. 13 feet

3. Grounding electrode conductors smaller than _____ AWG shall be in rigid metal conduit, intermediate metal conduit, rigid nonmetallic conduit, electrical metallic tubing, or cable armor.

    a. No. 10        b. No. 1/0        c. No. 6        d. No. 2

4. To ensure electrical continuity and the capacity to conduct safely any _____ current likely to be imposed, metal enclosures shall be effectively bonded.

    a. over        b. fault        c. hazardous        d. stray

5. When a circuit is protected by a 40A overcurrent device, the circuit's equipment grounding conductor must not be less than a _____.

    a. No. 8 copper        b. No. 10 aluminum        c. either a or b        d. neither a nor b

6. The voltage developed between the portable or mobile equipment frame and ground by the flow of maximum ground-fault current shall not exceed _____.

    a. 10 volts        b. 35 volts        c. 70 volts        d. 100 volts

7. Where rock bottom does not allow a ground rod to be driven 8' in depth, the rod can be _____.

    a. driven at an oblique angle not to exceed 45 degrees from the vertical

    b. driven at an angle up to 45 degrees

    c. buried in a trench that is 30" deep

    d. a or b

    e. a or c

    f. b or c

    g. a, b, and c

8.  In no case shall the grounded system conductor from the neutral point of a transformer or generator be smaller than _____.

    a. No. 6 AWG aluminum or copper-clad      b. No. 8 AWG copper      c. either a or b
    d. neither a nor b

9.  The minimum spacing between multiple electrodes shall not be less than _____ feet apart.

    a. 4      b. 5      c. 7      d. 6

10. An example of a ground-fault current path is an electrically conductive path between the earth and electrical equipment.

    a. True      b. False

11. When a copper service conductor is sized at 1000 kcmil, the grounding electrode conductor must be no less than _____.

    a. No. 3/0 copper      b. No. 250 aluminum      c. either a or b      d. neither a nor b

12. A bare conductor not smaller than a No. 4 AWG can serve as a grounding electrode when encased in concrete and at least _____ or more in length.

    a. 17'      b. 10'      c. 14'      d. 20'

13. Unless spliced by an irreversible compression type connector or exothermic welding, a grounding electrode conductor shall be in one continuous length without splice or joint.

    a. True      b. False

14. The equipment bonding jumper on the load side of the service overcurrent devices shall not be smaller than _____ AWG.

    a. No. 18      b. No. 16      c. No. 14      d. No. 12

15. A new branch circuit for a wall-mounted oven is required to be wired with a _____.

    a. multiconductor cable having three insulated conductors with an equipment ground

    b. multiconductor cable having two insulated conductors with an equipment ground

    c. either a or b

    d. neither a nor b

16. The frame of a portable generator is not required to be connected to a grounding electrode for a system supplied by a generator per NEC 250.52 when _____.

    I. the generator supplies only cord-and-plug connected equipment through a receptacle mounted on the generator

    II. the generator supplies only equipment mounted on the generator

    III. the normally non-current carrying metal parts of equipment and the equipment grounding terminals of the receptacle are connected to the generator frame.

    a. only II      b. only I      c. only III      d. only I and III
    e. only II and III      f. only I and II      g. I, II, and III      h. neither I, II, nor III

17. The equipment grounding conductor for secondary circuits of instrument transformers and for instrument cases shall not be smaller than No. _____ aluminum.

    a. 14      b. 12      c. 10      d. 8

18. Secondary circuits, per 411.5(A) and 680.23(A)(2) for lighting systems, must _____.

    a. be grounded      b. have overload protection      c. not be smaller than No. 18 AWG
    d. not be grounded

19. An approved means for externally connecting a copper or other corrosion-resistant bonding or grounding conductor to a grounded raceway or equipment is a No. 6 AWG copper conductor with one end bonded to a grounded nonflexible metallic raceway or equipment and with 6" or more of the other end made accessible on the outside wall.

    a. True      b. False

20. Electrical systems that are grounded shall be connected to earth in a manner that will limit the _____ imposed by lighting, line surges, or unintentional contact with higher-voltage lines.

    a. magnetic field      b. current      c. voltage      d. heat

21. The size of the grounded conductor shall be based on the total _____ of the parallel conductors when the service-entrance phase conductors are installed in parallel.

    a. cross-sectional area      b. diameter      c. circular mil area      d. volume

22. Where a DC system consists of a _____, the grounding electrode conductor must not be smaller than the neutral conductor.

    a. balancer winding with overcurrent protection
    b. 3-wire balance set
    c. both a and b
    d. either a or b
    e. none of the preceding

23. That portion of a grounding electrode conductor that is the sole connection to a grounding electrode shall not be required to be larger than _____.

    a. No. 6 AWG aluminum    b. No. 4 AWG copper    c. No. 6 AWG copper
    d. No. 4 AWG aluminum    e. a or b    f. c or d

24. If installed on the outside of a raceway or enclosure, the length of an equipment bonding jumper must not exceed _____ feet.

    a. 10    b. 6    c. 3    d. 5

25. For existing installations, the frames of household cooking appliances and clothes dryers are permitted to be connected to a grounded circuit conductor.

    a. True    b. False

26. An equipment grounding conductor must be _____.

    a. solid    b. bare    c. stranded    d. insulated
    e. covered    f. copper    g. aluminum    h. all choices apply

27. An onsite AC generator used as an alternate AC power source is not considered a separately derived system if the grounded conductor is solidly interconnected to a service-supplied system grounded conductor.

    a. True    b. False

28. A rod, pipe, or plate electrode that does not have a resistance to ground of _____ or less can be augmented by one additional electrode.

    a. 5Ω    b. 25Ω    c. 50Ω    d. 75Ω

29. Except where permitted, a grounded conductor shall not be reconnected on the load side of the service disconnecting means.

    a. True    b. False

30. The earth can be used as the sole equipment grounding conductor or effective ground-fault current path.

    a. True    b. False

31. For multiphase systems having one wire common to all phase conductors, the common conductor shall be grounded.

    a. True    b. False

32. Where rigid steel conduit is used as an grounding electrode, it must be at least 8 feet in length and be of a trade size no less than _____.
    a. 1/2                b. 1                    c. 3/4                d. 1 1/4

33. The grounded conductor shall not be smaller than 12.5 percent of the area of the largest service-entrance phase conductors when they are larger than _____ copper or _____ aluminum.
    a. 1250, 1100       b. 2000, 1750          c. 1500, 1750        d. 1100, 1750

34. In buildings of multiple occupancy where the metal water piping system is metallically isolated from all other occupancies by use of nonmetallic water piping, the metal water piping system for each occupancy shall be permitted to be bonded to the _____ terminal of the panelboard or switchboard enclosure supplying that occupancy.
    a. grounded
    b. equipment grounding
    c. grounding electrode
    d. all of the preceding

35. Where installed for the reduction of electrical noise on the grounding circuit, a receptacle in which the grounding terminal is purposely insulated from the receptacle mounting shall be permitted.
    a. True                b. False

36. An equipment grounding conductor shall be increased in size proportionately according to circular mil area when the size of a grounded conductor has been increased.
    a. True                b. False

37. _____ shall be used around concentric or eccentric knockouts that are punched or otherwise formed so as to impair the electrical connection to ground.
    a. An equipment grounding conductor        b. Grounding conductors
    c. Bonding jumpers                          d. Grounding pigtails

38. Where a multigrounded neutral system is used, the maximum distance between any two adjacent electrodes shall not be more than _____.
    a. 350'               b. 725'                c. 980'               d. 1300'

39. For a high-impedance, grounded neutral system, the grounding electrode conductor shall be connected at any point from the grounded side of the grounding impedance to the equipment grounding connection at the service equipment or first system disconnecting means.
    a. True              b. False

40. The grounding of electrical systems, _____, surge arresters, _____, and conductive non-current-carrying materials and equipment shall be installed and arranged in a manner that will prevent objectionable current.
    a. grounding grid, surge-protective devices
    b. circuit conductors, overcurrent devices
    c. surge-protective devices, lightning rods
    d. grounding grid, circuit conductors
    e. circuit conductors, surge-protective devices

41. A main bonding jumper must be a _____.
    a. bus          b. wire          c. screw          d. all choices apply

42. Plate electrodes shall expose not less than 2ft² of surface to exterior soil.
    a. True              b. False

43. The connection of a grounding electrode conductor or bonding jumper to a grounding electrode is not required to be accessible.
    a. True              b. False

44. When connected by a fixed wiring method, which piece of equipment is not required to be grounded?
    a. electric signs      b. elevators and cranes      c. pipe organs      d. motor frames
    e. switchboard frames, 2-wire DC effectively insulated from ground
    f. motion picture equipment

45. A 2-wire DC system supplying premises wiring and operating at greater than _____ volts but not greater than 300 volts shall be grounded.
    a. 30          b. 50          c. 100          d. 115

46. Ground rings are required to be buried at a depth that is no less than _____ from the earth's surface.
    a. 1.5'          b. 2.5'          c. 3'          d. 3.5'

47. Equipment grounding conductors that are insulated or covered shall have a continuous outer finish that is either green or green with one or more white stripes, except as permitted otherwise.

    a. True          b. False

48. Insulated or covered equipment grounding conductors larger than No. _____ AWG copper or aluminum shall be permitted at the time of installation to be permanently identified as an equipment grounding conductor at each end and at every point where the conductor is accessible.

    a. 4          b. 8          c. 10          d. 6

49. Where isolated from possible contact by a minimum of 18", a metal elbow that is installed in an underground installation of rigid nonmetallic conduit is not required to be connected to the equipment grounding conductor.

    a. True          b. False

50. To ensure good electrical continuity, coatings such as paint and enamel must be removed from threads and other contact surfaces.

    a. True          b. False

51. The grounding electrode used for service or feeder equipment shall be permitted as the grounding electrode for a separately derived system when the system originates in listed equipment for use as service equipment.

    a. True          b. False

52. Aluminum ground rods and underground gas metal piping systems can be used as grounding electrodes.

    a. True          b. False

53. Not more than _____ conductor(s) shall be connected to the grounding electrode by a single clamp or fitting unless listed for multiple conductors.

    a. 2          b. 3          c. 1          d. 4

54. Cord and plug motor operated equipment such as lawn mowers and hedge clippers are required to be grounded.

    a. True          b. False

55. A 22' ground ring encircling a building or structure not smaller than No. _____ bare copper can serve as a grounding electrode.

a. 1/0 AWG          b. 250 kcmil          c. 2 AWG          d. 6 AWG

56. Secondary circuits of current and potential instrument transformers shall be grounded where the primary windings are connected to circuits of _____ or more to ground.

a. 100  volts          b. 150  volts          c. 250  volts          d. 300 volts

57. The grounding impedance shall be installed between the grounding electrode conductor and the neutral point derived from a grounding transformer where a neutral point is unavailable.

a. True          b. False

58. An insulated or cover equipment grounding conductor larger than No. _____ AWG copper or aluminum shall be permitted, at the time of installation to be permanently identified as an equipment grounding conductor at each end and at every point where the conductor is accessible.

a. 4          b. 8          c. 10          d. 6

59. Standard locknuts or bushings are permitted to be the sole means for bonding to ensure electrical continuity.

a. True          b. False

60. An example of a premises wiring system permitted to be grounded is a _____ connection.

a. closed delta          b. 4W/3φ/ delta          c. 3W-open delta          d. corner-grounded delta

# ● ARTICLE 280–SURGE ARRESTERS, OVER 1kV

1.  The conductor used to connect the surge arrester to _____ and to a grounding connection point per NEC 280.21 shall not be any longer than necessary and shall avoid unnecessary bends.

    a. line     b. bus     c. equipment     d. a or b     e. a, b, and c     f. a, b, or c

2.  For ungrounded or unigrounded primary systems, the spark gap or listed device shall have a _____ breakdown voltage of at least the primary circuit voltage.

    a. 50VDC     b. unlimited     c. 60Hz     d. limited

3.  A _____ shall not be installed where the rating of the surge arrester is less than the maximum continuous phase-to-ground power frequency voltage available at the point of application.

    a. lightning arrester     b. surge protector     c. surge arrester     d. surge limiter

4.  The conductor between the surge arrester and the line and the surge arrester and the grounding connection shall not be smaller than No. _____ AWG copper.

    a. 8     b. 1/0     c. 6     d. 12

5.  Surge arresters can only be located outdoors.

    a. True     b. False

6.  The rating of a silicon carbide-type surge arrester shall be not less than _____ percent of the maximum continuous operating voltage available at the point of application.

    a. 80     b. 125     c. 135     d. 150

# ⊙ ARTICLE 285–SURGE-PROTECTIVE DEVICES (SPDs), 1kV OR LESS

1. The TVSS shall be connected on the load side of the _____ overcurrent device in a separately derived system.

   a. first               b. second               c. third               d. either a, b or c

2. Where used at a point on a circuit, the TVSS shall be connected to each _____ conductor.

   a. grounded          b. ungrounded          c. unigrounded          d. multigrounded

3. The grounded conductor and the grounding conductor shall be interconnected only by the normal operation of the TVSS during a surge.

   a. True               b. False

4. A TVSS shall not be used where circuits exceed 600 volts.

   a. True               b. False

# CHAPTER 3–WIRING METHODS AND MATERIAL

⦿ **ARTICLE 300–WIRING METHODS**

1.  The minimum thickness of a steel sleeve or clip must be no less than _____.
    a. 1/4"            b. 1/16"            c. 1/8"            d. 1/32"

2.  Raceways shall be provided with expansion fittings where necessary to compensate for _____ expansion and contraction.
    a. area            b. repeated         c. thermal         d. undetermined

3.  A direct buried cable rated for 27kV must be covered no less than _____.
    a. 18"             b. 24"              c. 30"             d. 36"

4.  The location of underground service conductors that are not encased in concrete yet are buried 18" or more below grade shall be identified by placing a _____ 12" above the underground installation.
    a. sensing device  b. warning ribbon   c. metal plate     d. remote detector

5.  A box or conduit body shall not be required where accessible fittings are used for straight-through splices in _____ sheathed cable.
    a. nonmetallic     b. armored          c. mineral-insulated    d. choices don't apply

6.  The "S" loop method is a recognized technique to protect direct buried cables or enclosed conductors from damage when subject to movement by settlement or frost.
    a. True            b. False

7.  Metal raceways, cable armor, and other metal enclosures for conductors must be metallically _____ between cabinets, boxes, and fittings.
    a. joined          b. continuous       c. bonded          d. accessible

8.  Raceways containing ungrounded conductors in sizes of _____ or larger shall be protected by a substantial fitting.
    a. No. 4 AWG       b. No. 2 AWG        c. No. 6 AWG       d. No. 8 AWG

9. In installations where walls are frequently washed, a _____ airspace must be maintained between walls and surrounding electrical equipment.
   a. 1"                 b. 1/2"                 c. 5/8"                 d. 1/4"

10. Unless modified by other articles, Article 300 covers wiring methods for all wiring installations.
    a. True                 b. False

11. When a raceway or cable type wiring method is installed through bored holes in joists, rafters, or wood members, holes shall be bored so that the edge of the hole is not less than _____ from the nearest edge of the wood member.
    a. 1"                 b. 1/2"                 c. 1 1/4"                 d. 3/4"

12. Cable trays and raceways containing electrical conductors must not contain pipes or tubing used for air, gas, or water.
    a. True                 b. False

13. Direct buried cables or conductors shall be permitted to be spliced or tapped without the use of splice boxes.
    a. True                 b. False

14. Unless otherwise permitted, all conductors of the same circuit, including grounded (neutral) conductors, equipment grounding conductors, and bonding conductors, shall be contained in the same _____.
    a. trench        b. raceway        c. cord        d. gutter        e. all of the preceding

15. At least _____ of free conductor must be maintained at each outlet, junction, and switch point for making splices or wiring connections.
    a. 3"                 b. 8"                 c. 12"                 d. 6"

16. Cables and direct-buried conductors emerging from grade shall be protected to a point at least _____ above finished grade.
    a. 2'                 b. 4'                 c. 6'                 d. 8'

17. Conductors installed in metal enclosures carrying alternating current must be arranged to avoid heating the surrounding ferrous metal by _____.
    a. capacitance        b. induction        c. eddy currents        d. exposure

18. All conductors rated 600 volts, nominal, or less shall have a(n) _____ rating equal to at least the maximum circuit voltage applied to any conductor within the enclosure, cable, or raceway.

    a. temperature          b. voltage              c. insulation           d. ampacity

19. Steel plates are required to protect both metallic and non-metallic raceway.

    a. True                 b. False

20. Match the correct Metric Designator with Trade Sizes and vice-versa:

    | Metric Designator | Trade Sizes |
    |---|---|
    | 12 | _____ |
    | _____ | 3/4 |
    | 35 | _____ |
    | _____ | 2 |
    | _____ | 3 1/2 |
    | 103 | _____ |
    | 155 | _____ |

21. The shortest distance measured between a point on the top surface of any direct-buried conductor or cable or other raceway and the top surface of the finished grade is the NEC definition used for the term _____.

    a. backfill             b. depth                c. cover                d. burial

22. Unless otherwise permitted, conductors of circuits rated over 600 volts are allowed to occupy the same equipment wiring enclosure.

    a. True                 b. False

23. Multiconductor cable rated over 600 volts, having individually shielded conductors, requires the minimum bending radius to be _____ times the diameter of the individually shielded conductors.

    a. 8                    b. 7                    c. 12                   d. 5

24. An underground feeder supplying a 24-volt irrigation circuit is required to be installed through a bored hole under a highway. How deep must the hole be bored below grade?

    a. 6"                   b. 30"                  c. 18"                  d. 24"

25. Conductors of circuits rated 600 volts, nominal, or less, whether AC and DC circuits, are permitted to occupy the same equipment wiring enclosure, cable, or raceway.

    a. True              b. False

26. Where treated with a flame-retardant saturant, a braid-covered insulated conductor must be stripped back at a distance no less than _____ at the conductor terminals for each kilovolt of the conductor-to-ground voltage of the circuit.

    a. 1"              b. 3/4"              c. 5"              d. 2"

27. A _____ must be used at the end of a conduit that terminates underground where conductors or cables emerge as a direct burial wiring method.

    a. compression fitting      b. rubber plug      c. bushing      d. PVC male adapter

28. For _____ branch circuits, the continuity of a grounded (neutral) conductor must not depend upon device connections (such as lampholders and receptacles) where the removal of such devices would interrupt continuity.

    a. bathroom              b. individual              c. ungrounded              d. multiwire

29. A No. 350 kcmil aluminum conductor installed in a vertical raceway must be supported when the installation exceeds _____.

    a. 85'              b. 95'              c. 120'              d. 135'

30. Chapter 3 shall be used for _____ volts, nominal, or less if not specified elsewhere in some other part of the chapter.

    a. 300              b. 480              c. 600              d. 1000

# ● ARTICLE 310–CONDUCTORS FOR GENERAL WIRING

1. Ungrounded conductors shall be clearly distinguishable from grounded and grounding conductors.

   a. True               b. False

2. Listed wire types designated with the suffix "2" are permitted to be used at a continuous 90°C operating temperature whether used in a wet or dry location.

   a. True               b. False

3. When considering load diversity, Table B.310.11 of Annex B must be referenced to determine adjustment factors for more than three current-carrying conductors.

   a. True               b. False

4. The ampacity of a No. 2/0 aluminum conductor rated for 75°C is _____, as per Table 310.16.

   a. 175A       b. 100A       c. 135A       d. 150A

5. The grounded conductor of a 2-wire 120- or 277-volt circuit is considered current carrying.

   a. True               b. False

6. Where operated above 2000 volts in permanent installations, solid dielectric insulated conductors shall _____.

   a. be shielded

   b. have ozone-resistant insulation

   c. either a or b

   d. both a and b

7. A conductor with letter type MTW is used for _____.

   a. motors and transformer wiring

   b. mobile homes and trailers

   c. machine tool wiring

   d. receiving and transmitting electronic mail

8.  If a conductor is rated for 75°C and placed in an environment where the ambient temperature exceeds 30°C, the conductor's ampacity requires being derated to prevent insulation damage.

    a. True              b. False

9.  A bare conductor must take on an equivalent temperature rating to that of the lowest temperature rating of adjacent insulated conductors when considering its ampacity.

    a. True              b. False

10. Excluding exceptions, all conductors, according to the NEC, are required to be insulated.

    a. True              b. False

11. The correction factors listed for Tables 310.16–20 must be applied when the allowable surrounding temperature is exceeded.

    a. True              b. False

12. Derating factors shall not apply for conductors installed in nipples of a length not exceeding _____.

    a. 36"              b. 18"              c. 48"              d. 24"

13. No. 300 kcmil aluminum service conductors can be used to supply a residential service panelboard rated for _____ amperes, according to Table 310.15(B)(6).

    a. 200              b. 300              c. 175              d. 250

14. A conductor with insulation type TBS is only manufactured up to a _____ in size.

    a. No. 4              b. No. 1              c. No. 1/0              d. No. 4/0

15. Parallel conductors in each phase, neutral, grounded circuit conductor or equipment grounding conductor shall _____.

    a. have the same insulation type

    b. be the same size in circular mil area

    c. have the same conductor material

    d. be terminated in the same manner

    e. be the same length

    f. all of the preceding

16. Where more than one calculated or tabulated ampacity could apply for a given circuit length, the _____ value shall be used.

    a. average                b. lowest                c. highest                d. choices do not apply

17. Insulated conductors and cables used in dry and damp locations shall have which of the following insulation types?

    a. PFA                b. THHN             c. THWN-2            d. FEP

    e. only a and d        f. only b and c        g. only a, b and d      h. a, b, c, and d

18. Insulated conductors and cables used in wet locations shall be moisture-impervious metal-sheathed.

    a. True                b. False

19. A 50' run of 3/4" EMT is installed 1" above the rooftop of a building. The surrounding temperature of the tubing is required to be increased by _____.

    a. 60°F           b. 14°C           c. 22°F           d. 40°C           e. not applicable

20. If installed in raceways, conductors of size No. _____ AWG or larger shall be stranded, where permitted.

    a. 4                b. 8                c. 10                d. 6

21. The suffix letter _____ indicates an assembly of two or more insulated conductors twisted spirally within an outer nonmetallic covering.

    a. T                b. M                c. S                d. D

22. When the ambient temperature is 86°F and no more than three current-carrying conductors are involved, the ampacities listed in Table 310.16 _____.

    a. must be increased

    b. must be decreased

    c. can remain as is

    d. none of the preceding

23. Theoretically, for a balanced 480/277 volt multiwire circuit feeding lighting loads consisting only of incandescent bulbs, _____ ampere(s) will flow across the neutral conductor.

    a. very few           b. 0                c. cannot determine        d. less than 1

24. The temperature rating of a conductor is the _____ temperature at any location along its length that a conductor can withstand over a prolonged time period without serious degradation.

    a. operating          b. ambient                c. rated              d. maximum

25. The minimum size conductor permitted to be used for an electrical system rated less than 2000 volts is either a No. _____ copper or No. _____ aluminum or copper-clad aluminum.

    a. 12, 10          b. 14, 12          c. 10, 12          d. 12, 14

26. The ampacity of a No. 3 RHW-2 copper conductor supported by a messenger is _____.

    a. 100A          b. 138A          c. 107A          d. 144A

27. The ambient temperature for the correction factors in Tables 310.16–20 is required to be increased when conductors are installed in nonmetallic raceway that is exposed to direct sunlight.

    a. True          b. False

28. All insulated cables and conductors shall be marked to indicate _____.

    a. the AWG size or circular mil area

    b. the proper type letter(s)

    c. the manufacturer's name for identification

    d. the maximum rated voltage

    e. all of the preceding

29. Where eight current-carrying conductors are installed in the same raceway, the ampacity of each conductor must be reduced by _____ percent.

    a. 80          b. 70          c. 50          d. 45

30. Whether copper, aluminum, or copper-clad aluminum, the phase neutral or grounded circuit conductor allowed to be run in parallel must be no less than a _____.

    a. No. 1 AWG          b. No. 250 kcmil          c. No. 1/0 AWG          d. No. 300 kcmil

# ● ARTICLE 312–CABINETS, CUTOUT BOXES, AND METER SOCKET ENCLOSURES

1. For removable terminals intended for only one wire, bending space at terminals shall be permitted to be reduced by a pre-determined value.

   a. True                b. False

2. Cabinet and cutout boxes shall be deep enough to allow the closing of the doors when _____ branch-circuit panelboard switches are in any position.

   a. 30A          b. 15A          c. 20A          d. 25A

3. When a raceway containing an ungrounded conductor enters a meter can or cabinet, an insulated fitting or bushing is required for conductors sized at No. _____ or larger.

   a. 1/0          b. 6          c. 4          d. 1

4. For cabinets and cutout boxes, there shall be an airspace of at least _____ between any live metal part, including live metal parts of enclosed fuses, and the door.

   a. 2"          b. 3/4"          c. 1"          d. 1.25"

5. All cabinets, cutout boxes, and meter socket enclosures installed in wet locations shall be weatherproof.

   a. True                b. False

6. If a cutout box or cabinet is constructed of sheet steel, the metal thickness shall not be less than _____ uncoated.

   a. 0.53"          b. 0.0053"          c. 0.053"          d. 5.3"

7. An enclosure used for switches or overcurrent devices shall not be used as a junction box unless adequate space is provided so that the conductors will not fill the wiring space at any cross section to more than 40 percent of the cross-sectional area of that space, and the _____ will not fill the wiring space at any cross section to more than 75 percent of the cross-sectional area of that space.

   a. taps          b. conductors          c. splices          d. all of the preceding

8. When a set of three parallel No. 250 kcmil conductors enters a service panelboard opposite the terminating lugs, the minimum wire-bending space is required to be _____.

   a. 6"          b. 8"          c. 10"          d. 12"

# ● ARTICLE 314–OUTLET, DEVICE, PULL, AND JUNCTION BOXES; CONDUIT BODIES; FITTINGS; AND HANDHOLE ENCLOSURES

1. Fixture studs and hickeys count as _____ conductor(s) each, based on the largest conductor present in the box.

   a. 2          b. 1          c. 0          d. no answer

2. A lighting fixture weighing more than _____ shall be supported independently of an outlet box unless the box is listed and marked for the maximum weight to be supported.

   a. 22 lbs          b. 65 lbs          c. 50 lbs          d. 35 lbs

3. The permanently marked description DANGER–HIGH VOLTAGE–KEEP OUT must be block type and at least _____ inches in height.

   a. 3/8          b. 15/16          c. 1/2          d. 1

4. The volume of a wiring enclosure shall be the total volume of an assembled section and, where used, the space provided by _____ that are marked with their volume.

   I. dome covers          II. plaster rings          III. extension rings

   a. only I and III          b. only II and III          c. only I and II          d. I, II, and III

5. In walls having a drywall surface, boxes shall be installed so that the front edge of the box will not be set back off the finished surface more than _____.

   a. 1/8"          b. 1/16"          c. 1/4"          d. 3/16"

6. When a raceway enters a conduit body opposite a removable cover, the distance from the entry to the cover is based on the size of the conduit body.

   a. True          b. False

7. Six No. 8 conductors is the maximum number of conductors that can be installed in a _____ square box.

   a. 4 11/16" x 1 1/4"          b. 4" x 2 1/8"          c. 4" x 1 1/4"          d. 4 11/16" x 2 1/8"

8. A device box wider than a single _____ device box, per Table 314.16(A), shall have double-volume allowances provided for each gang required for mounting.

   a. 2-gang          b. metallic          c. nonmetallic          d. FS          e. not applicable

9.  In completed installations, each box shall have a _____.

    a. faceplate

    b. cover

    c. lampholder

    d. fixture canopy

    e. all of the preceding

10. Where conduits or connectors require the use of locknuts or bushings to be connected to a box, round boxes shall not be used.

    a. True                    b. False

11. When a wooden brace is used for mounting a box, it shall have a cross section not less than _____.

    a. 2" x 4"           b. 1 1/2" x 2"           c. 2" x 3"           d. 1" x 2"

12. An outlet box designed to support a six-blade ceiling fan weighing 44 pounds must be marked to reflect the maximum number of conductors the box can contain.

    a. True                    b. False

13. Conduit bodies containing conductors of No. 6 and smaller must have a cross-sectional area at least twice that of the largest conduit to which they can be attached.

    a. True                    b. False

14. Sheet steel boxes not over 100 in$^3$ in size shall be made from steel not less than _____ thick.

    a. 6.25 inches

    b. 0.075 inches

    c. 0.625 inches

    d. none of the preceding

15. Plastic boxes shall be durably and legibly marked by their manufacturer with their volume.

    a. True                    b. False

16. Wire nuts and other type small fittings are required to be considered when determining the number of conductors allowed in a box.

    a. True                    b. False

17. Each coil of an unbroken conductor not less than 12" must be counted twice.
    a. True                    b. False

18. For a straight pull box containing circuits rated less than 600 volts, the length of the box must not be less than _____ times the trade size of the largest raceway.
    a. 10          b. 6                    c. 12                    d. 8

19. The listed volume for a No. 8 AWG conductor is _____.
    a. 1.25 in³          b. 2.5 in³                    c. 4.25 in³                    d. 3 in³

20. Splices, taps, or devices shall not be enclosed in conduit bodies such as capped and service-entrance elbows when conductors sized at No. _____ and smaller are used.
    a. 10          b. 8                    c. 4                    d. 6

21. A 3-gang metal device box contains two cable clamps, one #14/2 and one #12/3 romex cables with grounds, two single-pole switches, and one duplex receptacle. A No. 12 conductor terminates on the terminals of the switches, and a No. 14 conductor terminates on the terminal of the receptacle. What size box is needed?
    a. 25 in³          b. 29.5 in³                    c. 32.25 in³                    d. 35.75 in³
    e. answer not provided

22. Boxes intended to enclose flush devices shall have an internal depth of not less than _____.
    a. 15/16"          b. 1"                    c. 7/8"                    d. 3/4"

23. The maximum number of No. 10 conductors permitted in a 3" x 2" x 2 1/4" device box is _____.
    a. 10          b. 7                    c. 4                    d. 6

24. Where multiconductor type UF cable is used with a single gang box not larger than 4" x 2 1/4", the cable is not required to be secured to the box if the cable is fastened within _____ of the box.
    a. 10"          b. 8"                    c. 6"                    d. 4"

25. For an angle pull box containing circuits rated over 600 volts, the distance between each conductor or cable entry and the opposite wall must not be less than _____ times the diameter, over sheath, of the largest cable or conductor.
    a. 60          b. 36                    c. 28                    d. 42

26. Boxes, conduit bodies, and fittings installed in wet locations shall be listed for use in wet locations.

    a. True                 b. False

27. A box shall be supported from a multiconductor cord or cable in an approved manner that protects the conductors against strain.

    a. True                 b. False

28. The number of conductors allowed in a single gang FS box must be limited to a volume of _____.

    a. 10.75 in$^3$         b. 11.75 in$^3$         c. 13.5 in$^3$         d. 18.0 in$^3$

# ● ARTICLE 320–ARMORED CABLE: TYPE AC

1. Type AC cable shall have an armor of flexible metal tape and shall have an _____ strip of copper or aluminum in intimate contact with the armor for its entire length.

   a. equipment grounding

   b. internal bonding

   c. external bonding jumper

   d. all choices are acceptable

2. Type AC cable must be secured by staples or similar fittings at intervals not exceeding 4 1/2" and within 12" of every fitting, junction box, cabinet, and outlet box.

   a. True                    b. False

3. The use of Type AC cable is not permitted in _____.

   I. damp locations          II. wet locations          III. dry locations

   a. I and II          b. II and III          c. I and III          d. I, II, and III

4. The radius of the curve of the inner edge of any bend shall not be less than _____ times the diameter of Type AC cable.

   a. 9          b. 7          c. 5          d. 3

5. The ampacity of Type AC cable installed in thermal insulation shall be based on that of a _____ conductor.

   a. 75°C          b. 90°C          c. 60°C          d. 105°C

6. If Type AC cable is run across the top of floor joist in an attic that is not accessible by means of permanent stairs or ladder, the cable must be protected only within _____ of the nearest edge of a scuttle hole or attic entrance.

   a. 4'          b. 5'          c. 6'          d. 7'

7. Type AC armored cable is a fabricated assembly of insulated conductors in a flexible metallic enclosure.

   a. True                    b. False

# ● ARTICLE 322–FLAT CABLE ASSEMBLIES: TYPE FC

1.  Flat cable assemblies shall have conductors of _____ AWG special stranded copper wires.

    a. 12                 b. 8                 c. 14                 d. 10

2.  Fixture hangers installed with flat cable assemblies shall be identified for the use.

    a. True               b. False

3.  Where identified for use with Type FC cable, the final section opposite the grounded conductor section of a terminal block shall a _____ marking or other suitable designation.

    a. black

    b. red

    c. yellow

    d. purple

    e. none of the preceding

4.  Where flat cable assemblies are permitted, branch circuits that supply suitable tap devices for lighting, small appliances, or small power loads must be rated for not more than _____.

    a. 25A                b. 20A               c. 30A                d. 15A

5.  Tap devices for flat cable assemblies shall be rated at not less than _____.

    a. 25A                b. 15A               c. 10A                d. 20A

# ● ARTICLE 324–FLAT CONDUCTOR CABLE: TYPE FCC

1. Type FCC cable shall be listed for use with an FCC system and shall consist of three to five copper conductors, one of which shall be a(n) _____ conductor.

   a. grounded          b. shielded          c. equipment grounding          d. grounding electrode

2. Where Type FCC cable is used for individual branch circuits, the rating of the circuit must not exceed _____ amperes.

   a. 50          b. 15          c. 30          d. 45

3. All bare FCC cable ends shall be insulated and sealed against moisture and liquid spillage by using listed insulating ends.

   a. True          b. False

4. A complete wiring system for branch circuits that is designed for installation under _____ is called an FFC system.

   a. carpet          b. floors          c. ground          d. carpet squares

5. Along with the information required in NEC 310.11(A), FCC cable shall be clearly and durably marked to identify the _____.

   a. ampacity of the cable

   b. cable's maximum temperature rating

   c. conductor's material type

   d. all of the preceding

6. Type FCC cable consist of _____ or more flat copper conductors placed edge-to-edge and separated and enclosed within an insulating assembly.

   a. one          b. two          c. three          d. four

7. The maximum voltage between ungrounded and grounded conductors must not exceed _____ volts.

   a. 300          b. 150          c. 240          d. 115

8. Any portion of an FCC system with a height above floor level exceeding _____ shall be tapered or feathered at the edges to floor level.

   a. 9"          b. .5"          c. .09"          d. 1 5/8"

# ⦿ ARTICLE 326–INTEGRATED GAS SPACER CABLE: TYPE IGS

1. The ampacity of a No. 4250 kcmil IGS cable is _____.

   a. 357A              b. 206A              c. 519A              d. 491A

2. A run of Type IGS cable between pull boxes or terminations shall not contain more than the equivalent of four quarter bends, including those located immediately at the pull box or terminations.

   a. True              b. False

3. The actual inside diameter for a 4 conduit used for IGS cable is _____.

   a. 4.5"              b. 2.375"            c. 3.710"            d. 1.947"

4. Type IGS cable can be used as interior wiring or be exposed in contact with buildings.

   a. True              b. False

5. The nominal gas pressure for an IGS cable shall be _____ pounds per square inch gauge.

   a. 35                b. 20                c. 138               d. 97

6. The minimum bending radius for a 3" nonmetallic flexible conduit enclosing an IGS cable is _____.

   a. 45"               b. 24"               c. 60"               d. 35"

# ● ARTICLE 328–MEDIUM VOLTAGE CABLE: TYPE MV

1. Type MV cable is defined as a single or multiconductor solid dielectric insulated cable rated for _____ volts or higher.

   a. 1000          b. 550          c. 1750          d. 2001

2. Type MV cable shall be permitted for use on power systems rated up to 42 kV in wet locations.

   a. True          b. False

# ● ARTICLE 330–METAL-CLAD CABLE: TYPE MC

1. The ampacity for Nos. 18 and 16 Type MC cable must be in accordance with Table
   _____.

   a. 310.16            b. 310.60            c. 400.5            d. 402.5

2. Type MC cable containing four or fewer conductors, sized no larger than _____, shall
   be secured within 12 inches of every box, cabinet, fitting, or other cable termination.

   a. No. 14            b. No. 12            c. No. 10            d. No. 8

3. Type MC cable with a smooth sheath that has an external diameter of 3/4" requires a bending radius of _____ times the external diameter of the metallic sheath.

   a. 7            b. 10            c. 12            d. 15

4. Type MC cable cannot be used for direct burial installations.

   a. True            b. False

5. The minimum size conductors used for Type MC cable can be no smaller than No.
   _____ copper and No. _____ aluminum or copper-clad aluminum.

   a. 12, 14            b. 18, 12            c. 16, 18            d. 16, 14

6. Where single conductor Type MC cable is used, all phase conductors and neutral conductors when used shall be grouped together to _____ induced voltage on the sheath.

   a. maximize            b. eliminate            c. minimize            d. reduce

7. A factory assembly of one or more insulated circuit conductors with or without optical fiber
   members enclosed in an armor of interlocking metal tape, or smooth or corrugated metallic
   sheath is known as _____ cable.

   a. fiber optic            b. armored cable            c. metal-clad            d. metallic-sheathed

# ● Article 332–Mineral-Insulated, Metal-Sheathed Cable: Type MI

1. Type MI cable shall be of _____ with a resistance corresponding to standard AWG and kcmil sizes.

   a. nickel-coated copper

   b. solid copper

   c. nickel

   d. all of the preceding

2. Type MI cable that has an external diameter greater than 3/4" and not more than 1" requires a bending radius of _____ times the external diameter of the metallic sheath.

   a. 7          b. 10          c. 12          d. 15

3. Where the outer sheath of Type MI cable is made of steel, a separate _____ conductor shall be provided.

   a. bonding          b. grounded          c. equipment grounding          d. shielded

4. Where Type MI cable terminates, an end seal fitting shall be installed immediately after stripping to prevent the entrance of gas into the insulation.

   a. True          b. False

5. Type MI cable is only permitted to be used in designated classified locations.

   a. True          b. False

6. Type MI cable shall be securely supported at intervals not exceeding _____ feet.

   a. 2          b. 6          c. 8          d. 10

# ● Article 334–Nonmetallic-Sheathed Cable: Types NM, NMC, and NMS

1.  The ampacity of Types NM, NMC, and NMS cable must be that which is used for conductors having a _____ temperature rating.
    a. 194°F             b. 60°C                  c. 145°F                 d. 75°C

2.  The bending radius of Types NM, NMC, and NMS cables must not be less than six times the diameter of the cable.
    a. True              b. False

3.  Type NMS cable can be used as a hybrid cable to provide power, communication, and signaling needs.
    a. True              b. False

4.  Nonmetallic sheathed cable is only manufactured in size up to a _____.
    a. 350 kcmil         b. No. 1/0 AWG           c. No. 250 kcmil         d. No. 2 AWG

5.  An outlet box is not required for switches, outlets, and tap devices of insulating material for rewiring in existing buildings where Types NM, NMC, and NMS cable are concealed and fished.
    a. True              b. False

6.  For Type NMC cable, the letter "C" stands for _____.
    a. cable             b. ceiling rated         c. corrosion resistant   d. concrete approved

7.  Types NM, NMC, and NMS cable are only permitted to have a bare conductor for equipment-grounding purposes.
    a. True              b. False

8.  Where fished in access points that are concealed in finished buildings or structures, nonmetallic sheathed cable is not required to be secured.
    a. True              b. False

9.  Where nonmetallic sheathed cable is run at angles with joist in unfinished basements, it shall be permissible to secure cables not smaller than _____ conductors directly to the lower edges of the joists.
    a. No. 10/3      b. No. 8/3       c. No. 6/2       d. a and b      e. b and c      f. a and c

10. Types NM, NMC, and NMS cables are permitted to be used in _____ dwellings.

    I. one-family          II. two-family          III. multifamily

    a. only II and III     b. only I and II     c. only I and III          d. I, II, and III

11. When required to pass through a floor, nonmetallic sheathed cable is required to be enclosed in _____.

    a. rigid metal conduit extending at least 12" above the floor

    b. nonmetallic conduit extending at least 6" above and below the floor

    c. EMT extending at least 6" below the floor

    d. metal or Schedule 80 nonmetallic raceway extending at least 6" above the floor

12. Nonmetallic sheathed cable must be secured within _____ of every cabinet, outlet box, or fitting.

    a. 6"               b. 54"                    c. 12"                    d. 36"

# ● ARTICLE 336–POWER AND CONTROL TRAY CABLE: TYPE TC

1. The smallest insulated conductors of Type TC tray cable is No. 16 AWG copper and No. 14 AWG aluminum or copper-clad aluminum.

    a. True                    b. False

2. The bending radius for Type TC cable without metal shielding must not be less than _____ times the overall diameter for cables larger than 2" diameter.

    a. 4                b. 5                c. 6                d. 12

3. Type TC tray cable is permitted to be used _____.

    a. for Class I circuits

    b. for power, lighting, control and signal circuits

    c. in cable trays

    d. in wet locations

    e. all of the preceding

# ● ARTICLE 338–SERVICE-ENTRANCE CABLE: TYPES SE AND USE

1. Type USE cable can be used for underground service entrance installations.

    a. True                    b. False

2. Type SE or USE cable containing two or more conductors shall be permitted to have _____ conductor(s) uninsulated.

    a. 0                b. 1                c. 2                d. answer not provided

3. Type SE service-entrance cable is permitted for use where the insulated conductors are used for circuit wiring and the uninsulated conductor is used only for _____ purposes.

    a. bonding

    b. grounding

    c. equipment grounding

    d. all of the preceding

# ARTICLE 340–UNDERGROUND FEEDER AND BRANCH-CIRCUIT CABLE: TYPE UF

1. A No. _____ AWG is the maximum size conductor that can be used for underground feeder and branch circuit cable regardless of conductor type.

   a. 2                 b. 1                 c. 2/0                 d. 4/0

2. The ampacity of Type UF cable is limited to that which is used for conductors having a _____ temperature rating.

   a. 60°C              b. 90°C              c. 194°F              d. 75°C

3. Type UF is permitted to be used in designated areas of hazardous locations.

   a. True              b. False

4. Type UF cable must have an overall covering rated for moisture and flame retardant.

   a. True              b. False

# ● ARTICLE 342—INTERMEDIATE METAL CONDUIT: TYPE IMC

1.  Intermediate metal conduit must be supported at intervals not exceeding _____ feet.

    a. 5                  b. 10                  c. 15                  d. 20

2.  The maximum trade size IMC permitted by the NEC is _____.

    a. 2"                 b. 3"                  c. 4"                  d. 5"

3.  Aluminum fittings and enclosures are not allowed to be used with intermediate metal conduit.

    a. True               b. False

4.  IMC is permitted to be used as an equipment grounding conductor.

    a. True               b. False

5.  Where conduit is threaded in the field, a standard cutting die with a _____ shall be used.

    a. 1/2" taper/ft      b. taper of 1 in 16    c. 3/4" taper/ft       d.1" taper/ft$^2$

6.  Factory elbows, couplings, and associate fittings must be labeled.

    a. True               b. False

7.  Type IMC must not contain more than the equivalent of four quarter bends or _____ total.

    a. 270°               b. 360°                c. 180°                d. 425°

# ⦿ ARTICLE 344–RIGID METAL CONDUIT: TYPE RMC

1.  Where used in wet locations, bolts and screws associated with the installation of Type RMC must be made of material that is corrosion-resistant.
    a. True          b. False

2.  Running threads must not be used on conduit for connection at couplings.
    a. True          b. False

3.  Where multiple runs of rigid metal conduit are installed in trade sizes above 3", it is required to be supported every _____.
    b. 10'        b. 15'        c. 20'        d. 30'

4.  Type RMC can be purchased in trade size up to _____.
    a. 3"        b. 4"        c. 5"        d. 6"

# ⦿ ARTICLE 348–FLEXIBLE METAL CONDUIT: TYPE FMC

1.  When internal fittings are used with 3/8" flexible metal conduit, _____ No. 14 PGF copper conductors can be installed in the conduit.
    a. five        b. four        c. six        d. three

2.  Angle connectors used with Type FMC are allowed to be used in concealed locations.
    a. True          b. False

3.  A raceway of circular cross section made of helically wound, formed, interlocked metal strip is known as _____.
    a. electrical nonmetallic tubing       b. liquidtight flexible nonmetallic conduit
    c. flexible metallic tubing       d. flexible metal conduit

4.  Type FMC can be used for underground installations.
    a. True          b. False

# ⊙ ARTICLE 350–LIQUIDTIGHT FLEXIBLE METAL CONDUIT: TYPE LFMC

1. Where fished behind concealed walls, liquidtight flexible metal conduit is not required to be securely fastened.

   a. True            b. False

2. Where subject to physical damage, Type LFMC is not permitted to be used.

   a. True            b. False

3. Where suitable for direct burial, Type LFMC shall be so marked.

   a. True            b. False

# ● ARTICLE 352–RIGID POLYVINYL CHLORIDE CONDUIT: TYPE PVC

1. All cut ends of rigid nonmetallic conduit must be _____.

   a. reamed            b. grinded            c. trimmed            d. filed

2. Where there is a 50°F change in temperature, the length of rigid nonmetallic conduit will expand _____ in./100 ft.

   a. 2.03              b. 3.04               c. 4.06               d. 5.07

3. Type RNC of a 2 1/2" trade size is required to be supported every _____.

   a. 3'                b. 5'                 c. 6'                 d. 7'

4. Where used in wet locations, supports and straps associated with the installation of Type RNC must be made of material that is corrosion-resistant.

   a. True              b. False

5. The maximum trade size Type RNC permitted by the NEC is _____.

   a. 3"                b. 4"                 c. 5"                 d. 6"

6. When a grounded conductor is enclosed in rigid nonmetallic conduit and used to ground equipment, a separate equipment grounding conductor is not required.

   a. True              b. False

7. Unless listed otherwise, rigid nonmetallic conduit must not be installed in areas where the ambient temperature exceeds _____.

   a. 50°C              b. 60°C               c. 75°C               d. 105°C

# ⦿ Article 353–High Density Polyethylene Conduit: Type HDPE Conduit

1. High density polyethylene conduit is used for the installation of _____ conductors.

    a. copper          b. electrical          c. ungrounded          d. non-current carrying

2. Because Type HDPE conduit is non-conductive, equipment grounding conductors are not required to be enclosed.

    a. True          b. False

3. HDPE conduit is manufactured in sizes _____ up to _____.

    a. 1/2", 6"          b. 1", 5"          c. 3/4", 6"          d. 1/2", 4"

# ⦿ Article 354–Nonmetallic Underground Conduit with Conductors: Type NUCC

1. Type NUCC is not permitted to be used in designated locations in classified areas.

    a. True          b. False

2. A factory assembly of conductors or cables inside a nonmetallic, smooth wall conduit with a circular cross section is the definition for _____.

    a. liquidtight flexible nonmetallic conduit

    b. nonmetallic underground conduit with conductors

    c. underground service entrance cable

    d. liquidtight flexible tubing

3. Identification of conductors or cables used in the assembly of Type NUCC must be provided on a tag attached to _____ of the assembly.

    a. both ends          b. the external surface          c. one end          d. all choices apply

4. The minimum bending radius for 2 1/2" nonmetallic underground conduit with conductors is _____.

    a. 18"          b. 60"          c. 48"          d. 26"          e. choice not provided

# ● ARTICLE 355–REINFORCED THERMOSETTING RESIN CONDUIT: TYPE RTRC

1.  RTRC shall be clearly and durably marked at least every 10' as required in the first sentence of NEC 110.21.

    a. True          b. False

2.  RTRC is permitted to be installed in areas where the ambient temperature exceeds 60°C.

    a. True          b. False

# ● ARTICLE 356–LIQUIDTIGHT FLEXIBLE NONMETALLIC CONDUIT: TYPE LFNC

1.  Horizontal runs of Type LFNC must be securely fastened within 6" of termination points.

    a. True          b. False

2.  Type LFNC-B as a listed manufactured prewired assembly is permitted for use in exposed locations up to a _____ inch trade size.

    a. 3      b. 4      c. 1      d. 2

3.  FNMC is an alternative designation for _____.

    a. LFNC-A

    b. LFNC-B

    c. LFNC

    d. all of the preceding

4.  A raceway of circular cross section with a corrugated internal and external surface without integral reinforcement within the conduit wall is designated as _____.

    a. LFNC-C      b. FNMC      c. LFNC-B      d. LFNC-A

# ● ARTICLE 358–ELECTRICAL METALLIC TUBING: TYPE EMT

1.  Where required, electrical metallic tubing can be threaded.

    a. True                    b. False

2.  Aluminum Type EMT is permitted to be used _____.

    a. in direct contact with the earth

    b. in concrete

    c. in areas subject to severe corrosive influences

    d. all of the preceding

3.  When installed in wet locations, connectors used with electrical metallic tubing must be _____ for use in wet locations.

    a. marked          b. classified          c. listed          d. approved

4.  Type EMT must be securely fastened in place at least every _____ feet.

    a. 3          b. 6          c. 10          d. 12

5.  Electrical metallic tubing is usually made of steel with a protective coating of aluminum.

    a. True                    b. False

# ● ARTICLE 360–FLEXIBLE METALLIC TUBING: TYPE FMT

1.  In installations where 3/4" Type FMT may be infrequently flexed, the minimum radius for flexing used must be at least _____.

    a. 10"          b. 12.5"          c. 17.5"          d. 20"

2.  Flexible metallic tubing cannot be used where the length will exceed 10 feet.

    a. True                    b. False

3.  A raceway that is circular in cross section, flexible, metallic, and liquidtight without a nonmetallic jacket is defined as Type _____.

    a. LNMC          b. FNMC          c. FMT          d. NUCC

# ● ARTICLE 362–ELECTRICAL NONMETALLIC TUBING: TYPE ENT

1. ENT larger than _____ inches in trade size shall not be used.

   a. 3           b. 1 1/2              c. 2              d. 2 1/2

2. When equipment grounding is required, a separate equipment grounding conductor must be installed in ENT.

   a. True              b. False

3. Where the voltage exceeds 1000 volts, ENT is not permitted.

   a. True              b. False

4. Because electrical nonmetallic tubing is a pliable raceway, it can be bent by hand with a reasonable force requiring no assistance.

   a. True              b. False

5. Where ENT enters a box, fitting, or other enclosure, a bushing shall be provided.

   a. True              b. False

6. Lengths not exceeding _____ from a fixture terminal connection for tap connections to lighting fixtures shall be permitted without being secured when ENT is used.

   a. 3'          b. 5'                c. 6'             d. 7'

7. In buildings exceeding _____ floors above grade, ENT shall be concealed within walls having a thermal barrier that has at least a 15-minute finish rating.

   a. two         b. five              c. three          d. four

# ● ARTICLE 366–AUXILIARY GUTTERS

1.  Where installed in extremely cold weather, nonmetallic auxiliary gutters may become brittle.

    a. True                b. False

2.  Adequate provisions in gutters shall be made for the expansion and contraction of

    _____.

    a. insulated conductors        b. busbars           c. bare conductors           d. metal parts

3.  An auxiliary gutter should not extend a distance greater than _____ feet beyond the equipment that it supplements.

    a. 10               b. 20               c. 30               d. 40

4.  Nonmetallic auxiliary gutters installed outdoors where exposed to rain shall be marked "suitable for use in wet locations."

    a. True                b. False

5.  Nonmetallic gutters are required to be supported at intervals not to exceed _____ at each end or joint unless listed for other support intervals.

    a. 10'              b. 7'               c. 5'               d. 3'

6.  Conductors, including splices and taps, shall not fill an auxiliary gutter to more than _____ percent of its area.

    a. 31               b. 40               c. 53               d. 75

7.  Auxiliary gutters are allowed to contain switches and overcurrent devices.

    a. True                b. False

8.  The derating factors of NEC 310.15(B)(2)(a) must only be applied when the number of current-carrying conductors, including neutrals classified as current-carrying, exceeds

    _____.

    a. 30               b. 45               c. 60               d. 80

# ● ARTICLE 368–BUSWAYS

1. Where busways are reduced in ampacity in an industrial building, overcurrent protection is not required _____.

   a. if the busway has an ampacity at least equal to 1/3 the rating or setting of the overcurrent device next back on the line

   b. the length of the busway having the smaller ampacity does not exceed 50 feet

   c. the busway is free from contact with combustible material

   d. all of the preceding

2. Unbroken lengths of a busway are allowed to be installed through dry walls.

   a. True                    b. False

3. Where busways are installed behind access panels, the space behind access panels _____.

    I. must not be used for air-handling purposes

   II. must not be used for environmental air

   a. I

   b. II

   c. both I and II

   d. neither I nor II

4. Where a busway is used as a feeder, a plug-in device shall consist of an externally operable circuit breaker or fusible switch.

   a. True                    b. False

5. Lighting and trolley busways should not be installed less than 10 feet above the floor unless provided with a cover identified for the purpose.

   a. True                    b. False

6. Vapor seals are not required in a forced-cooled bus.

   a. True                    b. False

7. A grounded metal enclosure containing factory-mounted, bare, or insulated conductors, which are usually copper or aluminum bars, rods, or tubes, is called a(n) _____.

   a. auxiliary gutter        b. load center        c. cabinet        d. busway

8. For each bus of a busway rated over 600 volts, a permanent nameplate must be provided reflecting the _____.

   a. rated voltage and current

   b. frequency and 60Hz withstand voltage

   c. rated continuous current and impulse withstand voltage

   d. only a and c

   e. only b and c

   f. a, b and c

9. Busways shall be securely supported at intervals not exceeding _____.

   a. 6'                b. 5'                c. 4'                d. 3'

10. Where required, the neutral bus of a busway must be sized to carry harmonic currents.

    a. True                b. False

11. A busway shall be protected against overcurrent based on the allowable _____ of the busway.

    a. full-load current        b. calculated load        c. current rating        d. nominal voltage

## ● ARTICLE 370–CABLEBUS

1. The ampacity of conductors in cablebus rated over 600 volts must be in accordance with Tables _____.

    a. 310.16 and .17     b. 310.17 and .19     c. 310.69 and .70     d. 310.85 and .86

2. Cablebus is assembled to carry _____ current and to withstand the magnetic forces of such current.

    a. rated     b. calculated load     c. fault     d. continuous

3. The size of cablebus conductors must not be smaller than a No. _____ AWG.

    a. 4     b. 2     c. 1/0     d. 4/0

4. Cablebus must not be installed in hoistways or hazardous locations.

    a. True     b. False

5. Cablebus shall be securely supported at intervals not exceeding 10 feet for vertical runs.

    a. True     b. False

## ● ARTICLE 372–CELLULAR CONCRETE FLOOR RACEWAYS

1. So-called loop wiring shall not be considered to be a splice or tap in header access units or junction boxes.

    a. True     b. False

2. A header that is installed with a cellular concrete floor raceway must be installed in a straight line at a _____ to the cells.

    a. 22.5° angle     b. 45° angle     c. 60° angle     d. 90° angle

3. The combined cross-sectional area of all conductors or cables shall not exceed _____ percent of the cross-sectional area of the cell or header.

    a. 31     b. 40     c. 60     d. 75

4. A single enclosed tubular space in a floor made of precast cellular concrete slabs is defined as a cell.

    a. True     b. False

# ⦿ ARTICLE 374–CELLULAR METAL FLOOR RACEWAYS

1. For future use, a suitable number of markers shall be installed for locating cells.
   a. True              b. False

2. When an outlet from a cellular metal floor raceway is abandoned, the sections of circuit conductors supplying the outlet _____.
   a. can be spliced         b. shall be removed from the raceway      c. can be reinsulated
   d. all of the preceding      e. none of the preceding

3. Cellular metal floor raceways are not permitted to be installed where subject to corrosive vapor.
   a. True              b. False

# ⦿ ARTICLE 376–METAL WIREWAYS

1. When insulated conductors No. _____ AWG or larger are pulled through wireways, the distance between raceway and cable entries enclosing the same conductor shall not be less than that required for straight and angle pulls.
   a. 1              b. 4              c. 2              d. 1/0

2. Dead ends of metal wireways must be closed.
   a. True              b. False

3. Sheet metal troughs with hinged or removable covers for housing and protecting electric wires and cable and in which conductors are laid in place after the wireway has been installed as a complete system are defined as a _____.
   a. gutter         b. metal wireway         c. junction box         d. cabinet

4. Vertical runs of wireways shall be securely supported at intervals not exceeding _____ feet.
   a. 5              b. 10              c. 15              d. 20

5. Where installed in wet locations of a classified area, wireways shall be _____ for the purpose.
   a. approved         b. listed         c. identified         d. marked

# ● ARTICLE 378–NONMETALLIC WIREWAYS

1. The derating factors of NEC 310.15(B)(2)(a) shall be applicable to the current-carrying conductors up to and including the _____ percent fill required for nonmetallic wireways.

   a. 20             b. 40             c. 60             d. 75

2. Conductors, including splices and taps, shall not fill a nonmetallic wireway to more than _____ percent of its area.

   a. 31             b. 40             c. 53             d. 75

3. Nonmetallic wireways are _____.

   a. weatherproof    b. flame retardant    c. raintight    d. heat resistant

4. Access to conductors shall be maintained on both sides of a wall when unbroken nonmetallic wireway extensions are passed transversely through a wall.

   a. True           b. False

# ● ARTICLE 380–MULTIOUTLET ASSEMBLY

1. Where provisions are made for removing the cap or covers on all exposed portions of partitions, a metal multioutlet assembly is permitted to pass through a dry partition.

   a. True           b. False

2. A multioutlet assembly is not permitted where the voltage is 250 volts or more between conductors.

   a. True           b. False

# ● ARTICLE 382–NONMETALLIC EXTENSIONS

1. Where receptacle-type tap connectors are used with nonmetallic extensions, they must be of the locking type.
   a. True                      b. False

2. One or more nonmetallic extensions are permitted to be run in any direction from an existing outlet, but not within _____ inches from the floor.
   a. 6              b. 4              c. 8              d. 2

3. Nonmetallic extensions are permitted when the extension originates from an existing outlet on a 15- or 20-ampere branch circuit.
   a. True                      b. False

# ● ARTICLE 384–STRUT-TYPE CHANNEL RACEWAY

1. Covers of strut-type channel raceway must be either metallic or nonmetallic.
   a. True                      b. False

2. Splices and taps are permitted in strut-type channel raceways but shall not fill the raceway to more than _____ percent of its area at that point.
   a. 40              b. 60              c. 75              d. 90

3. Strut-type channel raceway is permitted to be used when the voltage is less than 600 volts.
   a. True                      b. False

# ● ARTICLE 386–SURFACE METAL RACEWAYS

1. Where covers and accessories of nonmetallic materials are used on surface metal raceways, they shall be _____ for such use.
   a. labeled            b. listed             c. approved            d. identified

2. Surface metal raceways are permitted to be installed in _____ locations.
   a. Class II, Division I     b. Class I, Division 2     c. both a and b     d. neither a nor b

# ● ARTICLE 388–SURFACE NONMETALLIC RACEWAYS

1. The number of conductors or cables installed in surface nonmetallic raceway can exceed the number for which the raceway was designed.

    a. True            b. False

2. Where combination surface nonmetallic raceways are used for both signaling and for lighting and power circuits, the different systems can be run in the same compartments if properly identified.

    a. True            b. False

# ● ARTICLE 390–UNDERFLOOR RACEWAYS

1. Underfloor raceways spaced less than 1" apart shall be covered with concrete to a depth of 1 1/2".

    a. True            b. False

2. The combined cross-sectional area of all conductors or cables in underfloor raceway shall not exceed _____ percent of the interior cross-sectional area of the raceway.

    a. 20         b. 30         c. 40         d. 60

3. The installation of underfloor raceways in office occupancies is permitted where _____.

    a. covered with linoleum or equivalent floor covering

    b. laid flush with the concrete floor

    c. either a or b

    d. both a and b

4. Flat-top raceway not over _____ inches in width shall have not less than _____ inches of concrete or wood above the raceway.

    a. 3/4, 4         b. 3, 1/4         c. 4, 3/4         d. 1/4, 3/4

5. Junction boxes used with underfloor raceways shall be leveled to the floor grade and sealed to prevent the free entrance of water or _____.

    a. dust         b. moisture         c. concrete         d. condensation

# • ARTICLE 392–CABLE TRAYS

1. Aluminum cable trays shall not be used as equipment grounding conductors for circuits rated with ground-fault protection above _____.

   a. 600A        b. 1000A        c. 1600A        d. 2000A

2. Nonmetallic cable tray shall be made of a flame-retardant material.

   a. True        b. False

3. The maximum allowable rung spacing for ladder cable tray shall be _____ when Nos. 1/0 through 4/0 AWG single-conductor cables are installed.

   a. 9"        b. 18"        c. 12"        d. 15"

4. A unit or assembly of units or sections and associated fittings, forming a structural system used to securely fasten or support cables and raceway, is defined as a(n) _____.

   a. wireway        b. cable tray system        c. auxiliary gutter        d. busway

5. When Nos. 1/0 through 4/0 single conductors are installed in ventilated trough cable trays, all single conductors shall be installed in a single layer.

   a. True        b. False

6. Where conductors are passed from cable tray to raceway(s), the transition must not exceed _____.

   a. 3'        b. 5'        c. 4'        d. 6'

7. Where exposed to direct rays of the sun, insulated conductors and jacketed cables installed in a cable tray shall be identified as being _____.

   a. sun retardant        b. heat resistant        c. sunlight resistant        d. ultra-ray reflective

# ⦿ ARTICLE 394–CONCEALED KNOB-AND-TUBE WIRING

1.  Concealed knob-and-tube wiring shall be soldered unless approved splicing devices are used.

    a. True                 b. False

2.  Where solid knobs are used, conductors shall be securely tied to them by tie wires having insulation _____ that of the conductor.

    a. higher than          b. equivalent to          c. double          d. thicker than

3.  A clearance of _____ inches shall be maintained between conductors of concealed knob-and-tube wiring.

    a. 3                    b. 5                      c. 2               d. 4

# ⦿ ARTICLE 396–MESSENGER-SUPPORTED WIRING

1.  Messenger supported wiring shall not be used in hoistways or where subject to physical damage.

    a. True                 b. False

2.  An exposed wiring support system using a messenger wire to support insulated conductors by factory-assembled aerial cable is recognized as _____.

    a. an aerial cable system      b. open-conductor wiring      c. messenger-supported wiring
    d. factory-assembled wiring

# ● ARTICLE 398–OPEN WIRING ON INSULATORS

1. Conductors smaller than No. 8 AWG shall be rigidly supported within _____ of a dead-end connection to a receptacle.

   a. 8"                  b. 10"                  c. 6"                  d. 12"

2. Open wiring on insulators are permitted inside agricultural establishments if rated for 600 volts nominal or less.

   a. True                  b. False

3. Where nails are used to mount knobs, they shall not be smaller than ten penny.

   a. True                  b. False

4. No. 8 and larger conductors installed across open spaces are permitted to be supported up to 15 feet apart if noncombustible, nonabsorbent insulating spacers are used at least every _____ feet to maintain at least 2.5 inches between conductors.

   a. 4.5                  b. 7                  c. 9                  d. 6.5

5. Conductors used for open wiring shall be separated at least _____ from metal raceways, piping, or other conducting material.

   a. 4"                  b. 1"                  c. 3"                  d. 2"

# CHAPTER 4–EQUIPMENT FOR GENERAL USE

## ● ARTICLE 400–FLEXIBLE CORDS AND CABLES

1.  Winding flexible cords with tape is a method for preventing pull on a cord.
    a. True                b. False

2.  Type TST cord shall be permitted in lengths not exceeding 8' where attached directly, or by means of a special type plug, to a portable appliance rated _____ watts or more.
    a. 30            b. 50            c. 85            d. 100

3.  Conductors used with portable cables rated over 600 volts shall be No. _____ AWG or larger and shall employ flexible strandings.
    a. 12            b. 10            c. 8            d. 6

4.  Type SJEO cord can contain up to _____ conductors.
    a. 3            b. 4            c. 5            d. 6

5.  What is the ampacity of a No. 4/4 Type SCE cable rated for 75°C when three of the conductors are current carrying?
    a. 125A            b. 84A            c. 101A            d. 114A

6.  Type ETT elevator cable is permitted to be used in _____ locations.
    a. Class I, Div 2
    b. Class II, Div 1
    c. Class I, Div 1
    d. Class I, Div II
    e. all of the preceding

7.  Cables rated over 2000 volts shall be shielded for the purpose of confining the voltage _____ to the insulation.
    a. drop
    b. stresses
    c. loss
    d. all of the preceding

8. Which cord is permitted to be used for room air conditioners?

   a. Type W          b. Type SPT-3          c. Type SP-1          d. Type SRD

9. For jacketed cords furnished with appliances, one conductor must have its insulation colored _____.

   a. blue          b. red          c. light blue          d. orange-red

10. Type E elevator cable can be used in classified locations.

    a. True          b. False

11. The nominal installation thickness for ground-check conductors of Type G-GC cables shall not be less than _____ mils for No. 8 AWG.

    a. 15          b. 30          c. 45          d. 60

12. Flexible cords and cables are permitted to be used above suspended ceilings.

    a. True          b. False

13. Type S cord is permitted for _____ usage.

    a. hard          b. not hard          c. extra hard          d. choices not applicable

14. A three-conductor No. 20 AWG cable has an ampacity of _____ when used as elevator cable.

    a. 5A          b. .5A          c. 2A          d. 6A

# ● ARTICLE 402–FIXTURE WIRES

1.  Thermoplastic insulation associated with fixture wires may be deformed at normal temperatures when subjected to _____.

    a. heat

    b. pressure

    c. stress

    d. all of the preceding

2.  A No. 12 AWG Type ZF fixture wire has an allowable ampacity of _____ amperes.

    a. 17　　　　　　 b. 23　　　　　　 c. 28　　　　　　 d. 30

3.  The insulation thickness for Type KFF-1 fixture wires is _____ mils.

    a. 3　　　　　　 b. 5.5　　　　　　 c. 8.4　　　　　　 d. 20

4.  Fixture wires are not allowed to be used as _____ conductors except as permitted elsewhere in the NEC.

    a. neutral　　　 b. equipment grounding　　　 c. branch-circuit　　　 d. power

5.  Heat-resistant thermo plastic covered fixture wire with flexible strands has a maximum operating temperature of _____°F.

    a. 90　　　　　　 b. 140　　　　　　 c. 194　　　　　　 d. 302

6.  Unless otherwise specified, fixture wires are rated for service at _____ volts nominal.

    a. 125　　　　　　 b. 250　　　　　　 c. 450　　　　　　 d. 600

7.  Fixture wires are not permitted to be smaller than No. 18 AWG.

    a. True　　　　　　 b. False

8.  The maximum operating voltage for Type RFH-1 insulated fixture wire is _____ volts.

    a. 50　　　　　　 b. 150　　　　　　 c. 300　　　　　　 d. 600

# ● ARTICLE 404–SWITCHES

1.  A general-use snap switch suitable only for use on AC circuits can be used for controlling motor loads not exceeding _____ percent of the ampere rating of the switch at its rated voltage.

    a. 135          b. 150          c. 125          d. answer not available

2.  A snap switch must not be ganged in an enclosure with similar devices unless the switches are arranged so that the voltage between adjacent devices does not exceed _____ volts.

    a. 120          b. 250          c. 300          d. 600

3.  Auxiliary contacts of a renewable type shall be provided on all knife switches rated 600 volts and designed for use in breaking current over _____.

    a. 30A          b. 80A          c. 150A          d. 200A

4.  Double-throw knife switches shall be permitted to be mounted so that the throw will be

    _____.

    a. horizontal

    b. vertical

    c. either a or b

    d. neither a nor b

5.  Plastic faceplates shall not be less than _____ in thickness.

    a. .010"          b. .020"          c. .030"          d. .040"

6.  When a switch or circuit breaker is permitted to disconnect a grounded circuit conductor, all circuit conductors are required to be disconnected _____.

    a. by a qualified person

    b. at the same time

    c. both a and b

    d. neither a nor b

7.  A fused switch, only where permitted, is allowed to have fuses connected in parallel.

    a. True          b. False

8. All switches and circuit breakers shall be installed so that the center of the grip of the operating handle of the switch or circuit breaker, when in its highest position, is not more than _____ above the floor or working platform.

   a. 6'6"          b. 6'7"                    c. 6'9"                    d. 7'

9. Alternating current specific-use snap switches rated for 347 volts shall be listed and only used for controlling _____.

   a. noninductive loads     b. dimmer switches     c. inductive loads     d. a and b
   e. b and c                f. a and c

10. Three and four-way switches shall be wired so that all switching is done only in the _____ circuit conductor.

    a. traveler            b. ungrounded              c. common terminal     d. load

11. A DC general-use snap switch can be used for inductive loads not exceeding _____ percent of the ampere rating of the switch at the applied voltage.

    a. 35              b. 50              c. 80              d. 125

12. A standard 300A circuit breaker rated for 600 volts cannot be used to interrupt current under load.

    a. True             b. False

13. Switches shall not be installed within wet locations in _____ spaces unless installed as a listed assembly.

    a. shower

    b. tub

    c. either a or b

    d. neither a nor b

14. Snap switches are considered effectively grounded if _____.

    a. mounted with metal screws to a metal box

    b. mounted to a nonmetallic box with integral means for connecting to an equipment grounding conductor

    c. an equipment grounding conductor or bond jumper is connected to an equipment termination of the snap switch.

    d. only a and c

    e. all of the preceding

# ● ARTICLE 406–RECEPTACLES, CORDS CONNECTORS, AND ATTACHMENT PLUGS (CAPS)

1. When mounted on portable and vehicle-mounted generators, the equipment grounding conductor contacts for receptacle and cord connectors are not required to have those contacts connected to an equipment grounding conductor.

   a. True                    b. False

2. A terminal for connection to the grounding pole of a receptacle must be designated by _____, or, if the terminal for the equipment grounding conductor is not visible, the conductor entrance hole must be marked with the word green or ground.

   a. a green-colored pressure wire barrel

   b. a similar green-colored connection device in the case of adapters

   c. a green-colored hexagonal head or shaped terminal screw or nut

   d. all of the preceding

3. Where intended for use to reduce electrical noise, receptacles incorporating an isolated grounding conductor connection shall be identified by a(n) _____ located on the face of the receptacle.

   a. red dot            b. blue square            c. green star            d. orange triangle

4. Receptacles and cord connectors shall not be rated less than 15 amperes, _____ volts.

   a. 125                b. 250                    c. either a or b          d. neither a nor b

5. For 15 and 20 ampere 125 or 250 volt receptacles installed outdoors in wet locations, a receptacle is required to be installed in a _____ enclosure.

   a. weatherproof       b. waterproof             c. rainproof             d. moistureproof

6. When a grounding terminal does not exist on a receptacle, a nongrounding-type receptacle shall be permitted to be replaced with _____.

   a. a ground-fault circuit interrupter receptacle

   b. another nongrounding-type receptacle

   c. a grounding-type receptacle where supplied through a ground-fault circuit interrupter

   d. only b and c       e. only a and c           f. all of the preceding

7. Receptacles rated 20 amperes or less and designed for the direct connection of aluminum conductors shall be marked CU/AL.

   a. True                    b. False

# ⦿ ARTICLE 408–SWITCHBOARDS AND PANELBOARDS

1.  Where supplied from the secondary side of a transformer, a panelbaord must be protected by the overcurrent protection provided on the secondary side of the transformer where the protection is in accordance with NEC 240.21(C)(1).

    a. True                b. False

2.  The minimum spacing required between the bottom of a panelboard enclosure and noninsulated busbars is _____ inches.

    a. 8            b. 10            c. 5            d. 3

3.  The _____ phase shall be that phase having the higher voltage to ground on 3-phase, 4-wire, delta-connected systems.

    a. C            b. A            c. B            d. either phase is acceptable

4.  For other than a totally enclosed switchboard, a space not less than _____ feet shall be provided between the top of the switchboard and any combustible ceiling.

    a. 3            b. 4            c. 5            d. 6

5.  For existing panelboards, individual protection shall not be required for a panelboard used as service equipment for a commercial occupancy.

    a. True                b. False

6.  The phase arrangement for a 3-phase bus in a panelboard shall be A, B and C from _____.

    a. top to bottom

    b. front to back

    c. left to right

    d. all of the preceding

7.  A panelboard must be protected by an overcurrent device located on the supply side of the panelboard, per NEC 408.30.

    a. True                b. False

8.  An insulated conductor used within a switchboard shall be listed and flame retardant.

    a. True                b. False

9. Each grounded and equipment grounding conductors shall terminate within a panelboard in an individual terminal.

   a. True                 b. False

10. Article 408 also covers battery-charging panels supplied from light or power circuits.

    a. True                 b. False

11. Delta circuit breakers are not allowed to be used in panelboards.

    a. True                 b. False

12. Panelboards equipped with snap switches rated at _____ or less shall have overcurrent protection of _____ or less.

    a. 15A, 100A        b. 20A, 150A          c. 30A, 200A          d. 15A, 200A

13. Where rated for not over 250 volts, the minimum spacing between live parts to ground in a switchboard must be _____.

    a. 1/2"              b. 5/8"               c. 3/4"               d. 1'

## ● ARTICLE 409–INDUSTRIAL CONTROL PANELS

1. A main bonding jumper is required for an industrial control panel when connecting a grounded conductor on the load side to a grounding lug.

   a. True                 b. False

2. Multisection industrial control panels shall be bonded together with a(n) _____.

   a. bonding jumper

   b. equipment grounding conductor

   c. grounded conductor

   d. grounding strap

3. To determine the ampacity of supply conductors for an industrial control panel, the full-load current rating of all resistance heating loads must be increased by _____ percent.

   a. 110              b. 125                c. 140                d. 175

# ● ARTICLE 410–LUMINAIRES, LAMPHOLDERS, AND LAMPS

1. A handhole is not required for a metal pole supporting a light fixture that is _____ feet or less in height above grade that is provided with a hinged base.

   a. 20              b. 16              c. 12              d. 10

2. Tubing used as arms and stems on light fixtures with cut threads shall have a thickness not less than _____.

   a. .10"            b. .20"            c. .30"            d. .04"

3. The terminals of an electric-discharge lamp shall be considered as energized where any lamp terminal is connected to a circuit of over 120 volts.

   a. True            b. False

4. Track lighting is defined as a manufactured assembly that is designed to support and energize luminaries that are capable of being readily repositioned on track where its length can be altered by addition or subtraction of sections of track.

   a. True            b. False

5. Lampholders installed in wet or damp location shall be of the _____ type.

   a. rainproof       b. moistureproof   c. weatherproof    d. waterproof

6. Flexible cord shall be of the hard-service type, having conductors not smaller than the branch-circuit conductors, having ampacity at least equal to the _____, and having an equipment grounding conductor.

   a. rated amperes   b. branch-circuit overcurrent device   c. 60°C temperature rating
   d. calculated load

7. Lighting track conductors shall be a minimum of No. _____ AWG or equal and shall be copper.

   a. 16              b. 18              c. 14              d. 12

8. Unless identified for through-wiring, branch-circuit wiring should not pass through an outlet box that is an integral part of a luminaire.

   a. True            b. False

9. A minimum _____ spacing must be maintained between exposed live parts and the mounting plane of porcelain fixtures.

   a. 1"              b. 3/4"              c. 1/2"              d. 1/4"

10. Cleat-type lampholders located at 10' above the floor shall be permitted to have exposed terminals.

    a. True              b. False

11. An autotransformer that is used to raise the voltage to more than 300 volts, as part of a ballast for supplying lighting units, shall be supplied only by a(n) _____ system.

    a. ungrounded        b. grounded          c. either a or b       d. both a and b

12. Metal canopies supporting lampholders exceeding 8 lbs shall not be less than .010" in thickness.

    a. True              b. False

13. A ballast in a fluorescent exit fixture shall not have _____ protection.

    a. overload          b. short circuit     c. thermal             d. overcurrent

14. A 2000 watt incandescent lamp employed for general use on lighting branch circuits requires a _____ base.

    a. medium

    b. standard

    c. mogul

    d. none of the preceding

15. Transformers shall have a secondary short circuit rating of not more than _____ mA if the open-circuit voltage rating is 7500 volts or less.

    a. 75                b. 150               c. 300                 d. 450

16. A _____ fixture shall be permitted to be installed in a closet.

    a. recessed fluorescent

    b. completely enclosed surface-mounted incandescent

    c. surface-mounted fluorescent

    d. completely enclosed recessed incandescent

    e. only a and c

    f. a, b, c, and d

17. Electric-discharge lighting luminaires provided with mogul-base, screw-shell lampholders shall be permitted to be connected to branch circuits of _____ amperes or less.

    a. 20                b. 60                c. 30                d. 50

18. Where a luminaire is recessed in fire-resistant material in a building of fire-construction, a temperature not higher than _____ shall be considered acceptable, if marked for such use.

    a. 90°C              b. 150°F             c. 194°C             d. 302°F

19. The volume bounded by the sides and back closet walls and planes extending from the closet floor vertically to a height of 6 feet is defined as _____.

    a. a closet          b. storage space     c. closet storage space     d. closet space

20. Light fixtures installed in wet locations shall be marked "Suitable for _____ Locations."

    a. Outdoor      b. Wet      c. Soggy      d. Moist      e. all choices apply

21. The grounded conductor, where connected to a screw-shell lampholder, shall be

    _____.

    a. grounded      b. connected to the screw shell      c. labeled      d. terminated

22. Article 410 covers light fixtures, decorative lighting products, and equipment forming part of lighting installations.

    a. True              b. False

23. Fixtures and lampholders that weigh more than _____ or exceed 16" in any dimension shall not be supported by the screw shell of a lampholder.

    a. 10 lbs            b. 8 lbs             c. 12 lbs            d. 6 lbs

24. Equipment that has an open-circuit voltage exceeding 1000 volts shall not be installed in or on _____ dwellings.

    a. single-family

    b. two-family

    c. multifamily

    d. all of the preceding

25. Unless an individual switch is provided for light fixtures located over combustible material, lampholders shall be located at least _____ above the floor.

    a. 7'                b. 8'                c. 8.5'              d. 9'

26. When longer than 2', pendant conductors shall be twisted together where not cabled in a listed assembly.

    a. True                    b. False

27. Auxiliary equipment for electric discharge lamps shall be enclosed in combustible cases and treated as sources of _____.

    a. energy          b. heat          c. thermal protection          d. fire retardants

28. Branch-circuit conductors within _____ of a ballast shall have an insulation temperature rating not lower than 90°C.

    a. 3"              b. 2"              c. 4"              d. 1"

29. Track lighting is a manufactured assembly designed to support and energize luminaires that are capable of being readily repositioned on the track. Its length can be altered by the addition or subtraction of sections.

    a. True                    b. False

30. Ceiling-suspended fans shall not be located within a zone measured _____ feet horizontally and _____ feet vertically from the top of the bathtub rim or shower stall threshold.

    a. 8, 3           b. 2, 7           c. 3, 8           d. 4, 6

31. Pendant conductors, unless part of decorative lighting assemblies, shall not be smaller than _____ AWG for intermediate or candelabra-base lampholders.

    a. 16             b. 14             c. 18             d. 12

32. Combustible low-density cellulose fiberboard includes sheet, panels, and tiles that have a density of _____ or less.

    a. 2 lb/ft$^2$

    b. 3 lb/in$^3$

    c. 32 lb/ft$^3$

    d. 2 kg/ft$^3$

    e. none of the preceding

33. Luminaire studs that are not a part of outlet boxes, hickeys, tripods, and crowfeet must be made of steel, malleable iron, or other material suitable for the application.

    a. True                    b. False

34. Where double-pole switched lampholders are used, the switching device of the lampholder must disconnect at least one ungrounded conductor of the circuit supplied.

    a. True                    b. False

35. _____ conductors shall be used for wiring on fixture chains and on other movable or flexible parts.

    a. Stranded          b. Insulated          c. Flexible          d. Individual

36. Heavy-duty lighting track is lighting track identified for use exceeding _____ amperes.

    a. 15          b. 20          c. 25          d. 30

37. Wall-mounted recessed fluorescent fixtures are permitted to be installed in a closet if there is a minimum clearance of _____ inches maintained between the fixture and the nearest point of a storage space.

    a. 6          b. 12          c. 18          d. 24

38. Portable handlamps shall not be required to be grounded where supplied through an isolating transformer with an ungrounded secondary of not over _____ volts.

    a. 30          b. 50          c. 100          d. 125

39. A fixture requiring supply wire rated higher than 60°C shall be marked with the minimum supply wire temperature rating.

    a. True                    b. False

40. Luminaires shall be constructed with shades or guards so that combustible materials are not subjected to temperature in excess of 90°F.

    a. True                    b. False

41. Tap conductors that are permitted to be run from the fixture terminal connection to an outlet box shall not be more than _____ in length.

    a. 1'          b. 1.5'          c. 4'          d. 6'

# ◉ ARTICLE 411–LIGHTING SYSTEMS OPERATING AT 30 VOLTS OR LESS

1.  Lighting systems operating at 30 volts or less shall be supplied from a maximum _____ ampere branch circuit.

    a. 15                  b. 6                  c. 20                  d. 3

2.  Bare conductors used with a 24 volt lighting system shall not be installed less than _____ above the finished floor, unless specifically listed for a lower installation height.

    a. 7'                  b. 8'                  c. 9'                  d. 10'

3.  A lighting system shall not be installed within 5 feet of pools, spas, fountains, or similar locations.

    a. True                b. False

# ⦿ ARTICLE 422–APPLIANCES

1. The disconnecting means for a 1/4 hp motor can be a _____ if it is within the appliance or is capable of being locked in the open position.

   a. circuit breaker　　b. branch circuit switch　　c. either a or b　　d. neither a nor b

2. A household-type appliance with surface heating elements having a maximum demand of more than _____ amperes computed in accordance with Table 220.5 shall have its power supply subdivided into two or more circuits, each of which shall be provided with overcurrent protection rated at not over _____ amperes.

   a. 50, 60

   b. 50, 40

   c. 60, 30

   d. none of the preceding

3. A six-blade ceiling fan weighing 44 pounds must be supported independently of an outlet box.

   a. True　　　　　b. False

4. Steam boilers employing resistance-type immersion electric heating elements contained in a listed instantaneous water heater shall be permitted to be subdivided into circuits not exceeding _____ amperes.

   a. 60　　　　b. 80　　　　c. 100　　　　d. 120

5. Where the front drawer of an electric household range can be removed, the plug and receptacle supplying the range can serve as the disconnecting means.

   a. True　　　　　b. False

6. Screw-shell lampholders shall not be used with infrared lamps rated over _____ watts.

   a. 1000　　　　b. 300　　　　c. 1200　　　　d. 550

7. The nameplate of a single-phase, 4350 watt water heater rated for 230 volts is not marked with the rating of the appliance over current protection; therefore, the overcurrent protection must be sized at no more than _____ percent of the appliance rated current.

   a. 125　　　　b. 150　　　　c. 175　　　　d. 200

8. All heating elements that are rated over one ampere, replaceable in the field, and a part of an appliance shall be legibly marked with the ratings _____.

    a. in volts and watts

    b. with the manufacturer's part number

    c. in volts and amperes

    d. only a or c

    e. either a, b, or c

9. The length of an electrically operated kitchen waste disposer flexible cord shall not be less than _____ and not over _____.

    a. 18", 36"          b. 12", 24"          c. 24", 48"          d. 30", 42"

10. The branch-circuit rating of an appliance that is continuously loaded, other than a motor-operated appliance, shall not be less than _____ percent of the marked rating.

    a. 125          b. 150          c. 175          d. 225

11. Central heating equipment other than fixed electric space-heating equipment shall be supplied by an individual branch circuit.

    a. True          b. False

12. Cord and plug-connected portable free-standing hydromassage units and hand-held dryers shall be constructed to provide protection for personnel against _____ when immersed while in the on or off position.

    a. shock

    b. injury

    c. electrocution

    d. all of the preceding

13. All single-phase cord and plug connected high-pressure spray washing machines rated at 250 volts or less shall be provided with factory-installed _____ protection for personnel.

    a. overcurrent

    b. overload

    c. ground-fault circuit-interrupter

    d. supplemental

14. Branch circuit conductors supplying a fixed storage-type water heater that has a capacity not exceeding 120 gallons shall have a rating not less than _____ percent of the nameplate rating of the water heater.

    a. 80

    b. 135

    c. 150

    d. none of the preceding

15. Electrically heated smoothing irons shall be equipped with an identified _____ means.

    a. temperature-limiting        b. disconnecting        c. current-limiting

    d. heat-sensing

16. The length of a flexible cord used for either a dishwasher or trash compactor must be 3 feet to _____ feet measured from the face of the attachment plug to the plane of the rear of the appliance.

    a. 7                b. 6                c. 5                d. 4

# ARTICLE 424–FIXED ELECTRIC SPACE-HEATING EQUIPMENT

1.  Electric space-heating cable conductors located above thermal insulation having a thickness of _____ inches do not require temperature corrections.
    a. 6              b. 12              c. 2              d. 8

2.  A device to open all ungrounded conductors supplying heating panels shall function when a low resistance line to line fault occurs.
    a. True              b. False

3.  Means shall be provided to ensure that the fan circuit is _____ when any heater circuit is energized.
    a. disconnected        b. energized        c. identified        d. operable

4.  Cable heating elements must maintain a clearance not less than _____ from ventilating openings, recessed luminaires, and their trims and other such openings in room surfaces.
    a. 2"              b. 3"              c. 4"              d. 5"

5.  The size of branch-circuit conductors and over current protective devices for electrode-type boilers shall be calculated on the basis of 125 percent of the _____.
    a. total load
    b. total load including motors
    c. motor loads
    d. total load excluding motors

6.  Heating cables shall be furnished complete with factory-assembled nonheating leads at least _____ feet in length.
    a. 4              b. 5              c. 6              d. 7

7.  Ground-fault circuit-interrupter protection for personnel shall be provided for electrically heated floors in _____.
    a. hydromassage bathtub locations
    b. bathrooms
    c. either a or b
    d. both a and b

8. Resistance-type heating elements in electric space-heating equipment shall be protected at not more than _____.

    a. 175 percent of the nameplate current

    b. 60A

    c. 50A

    d. 125 percent of the branch circuit conductors

9. A complete assembly provided with a junction box or a length of flexible conduit for connection to a branch circuit is defined as a _____.

    a. wiring enclosure     b. heating panel          c. termination kit          d. splice box

10. Branch circuits supplying two or more outlets for fixed electric space heating equipment shall be rated _____ amperes.

    a. 30              b. 25              c. 20              d. 15              e. all choices apply

11. Means shall be provided to disconnect the _____ of all fixed electric space-heating equipment from all ungrounded conductors.

    a. motor controller(s)          b. supplementary over current protection          c. heater
    d. all choices apply

12. Heating panels or heating panel sets shall not exceed _____ of heated area.

    a. 35 watts/m²          b. 335 watts/ft²          c. 353 ft/watts²          d. 33 watt/ft²

13. To ensure an even distribution of air over the face of a heater, heaters installed with 4' of heat pump outlets may require pressure plates on the inlet side of the duct heater.

    a. True              b. False

14. The ampacity of field-wiring conductors between the heater and the supplementary overcurrent protective devices shall be sized at not less than 125 percent of the load served when the heater is rated for 50kW or more.

    a. True              b. False

15. Heating cable installed in masonry floors shall not exceed _____ watts/linear foot.

    a. 45              b. 25.5              c. 37              d. 16.5

16. Listed baseboard heaters include instructions that may not permit their installation below _____ outlets.

    a. switch              b. receptacle              c. power              d. communication

17. Where within sight from fixed electric space-heating equipment without a motor rated over 1/8 hp, a branch circuit switch or circuit breaker is permitted to serve as the disconnecting means.

    a. True                 b. False

18. Match the following heating cable lead wires per color identification to indicate its correct circuit voltage:

    a. red            _____ 120 volt

    b. orange         _____ 208 volt

    c. yellow         _____ 240 volt

    d. blue           _____ 277 volt

    e. brown          _____ 480 volt

19. Adjacent runs of heating cable not exceeding 9 watts/ft shall not be installed less than 1.5" on center.

    a. True                 b. False

20. The rating of an overcurrent protective device and the ampacity of branch circuit conductors supplying fixed electric space-heating equipment consisting of resistance elements, whether equipped with a motor or not, shall not be less than _____ percent of the total load of the motors and the heaters.

    a. 225            b. 150            c. 175            d. 125

# ● ARTICLE 426–FIXED OUTDOOR ELECTRIC DEICING AND SNOW-MELTING EQUIPMENT

1. All but _____ inches of nonheating leads of Type TW and other approved types not having a grounding sheath shall be enclosed in raceway within asphalt or masonry.

    a. 1 to 6          b. 3 to 4          c. 2 to 5          d. 3

2. Heating cable and tape are examples of resistance heaters.

    a. True          b. False

3. A(n) _____ transformer with a grounded shield between the primary and secondary windings shall be used to isolate the distribution system from the heating system.

    a. multi-tap          b. isolation          c. dual-winding          d. single-phase

4. Where readily accessible to the user of the equipment, the branch circuit switch or circuit breaker for fixed electric heating equipment shall be permitted to serve as the disconnecting means.

    a. True          b. False

5. Heating panels used for the purpose of melting snow must not exceed _____ of the heated area.

    a. 1300 ft/watts$^2$          b. 120 ft/watts$^2$          c. 1300 m/watts$^2$          d. 120 watts/ft$^2$

6. An impedance heating system that is operating at a voltage greater than 30 but not more than _____ shall be grounded at a designated point(s).

    a. 60          b. 80          c. 90          d. 95

7. A system in which heat is generated on the inner surface of a ferromagnetic envelope embedded in or fastened to the surface to be heated is defined as a(n) _____.

    a. heating system

    b. skin-effect heating system

    c. impedance heating system

    d. resistance heating system

8. Splices and terminations at the end of nonheating leads, other than the heating element end, shall be installed in a box or fitting.

    a. True          b. False

9. The ampacity of branch circuit conductors for snow melting equipment shall not be less than _____ percent of the total load of the heaters.

    a. 80                b. 110                c. 125                d. 140

10. Unless provisions are made for expansion and contraction, _____ shall not be installed if they bridge expansion joints.

    a. units              b. cables              c. panels              d. a and b

    e. b and c            f. a and c             g. a, b, and c

11. External surfaces of outdoor electric deicing and snow-melting equipment that operate at temperatures exceeding 60°F shall be physically guarded to protect against contact by personnel in the area.

    a. True               b. False

# ● ARTICLE 427–FIXED ELECTRIC HEATING EQUIPMENT FOR PIPELINES AND VESSELS

1. Not less than 6" of cold leads shall be provided within a junction box.

    a. True               b. False

2. Pull boxes used for skin-effect heating shall be of _____ construction when used for outdoor installations.

    a. weatherproof       b. watertight          c. rainproof           d. raintight

3. The presence of electrically heated pipelines, vessels, or both shall be evident by the posting of appropriate caution signs or markings not exceeding _____ along the pipeline or vessel.

    a. 25'                b. moderate            c. 18'                 d. 20'

4. A pipeline is described as a length of pipe including pumps, valves, flanges, control devices, strainers, and/or similar equipment for conveying fluids.

    a. True               b. False

5. The ampacity of the conductors connected to the secondary of the transformer shall be at least _____ percent of the total load of electric heating equipment supplying a vessel.

    a. 30                 b. 65                  c. 80                  d. not applicable

6. The ampacity of branch circuit conductors and the rating of overcurrent protection for fixed electric heating equipment for pipelines and vessels shall not be less than _____ percent of the total load of the heaters.

    a. 125                b. 110                 c. 150                 d. 140

# ● ARTICLE 430–MOTORS, MOTOR CIRCUITS, AND CONTROLLERS

1. When a motor terminal housing encloses rigidly mounted motor terminals, the minimum spacing between the line terminals and other uninsulated metal parts for a motor rated for 460 volts is _____.

   a. 1/8"          b. 1/4"                    c. 3/8"                    d. 1/2"

2. Conductors supplying more than one motor or a motor(s) and other load(s), shall have an ampacity _____.

   a. not less than the sum of all motors and other loads

   b. not less than 125 percent of the full-load current of the highest rated motor plus the sum of the full-load current ratings of all other motors in the group

   c. not less the 125 percent of the full-load current of all motors rated less than the highest rated motor

   d. not less than 125 percent of the full-load current of the highest rated motor plus the sum of the full-load current ratings of all other motors in the group plus the ampacity of other loads

3. For motors connected to a branch circuit by means of an attachment plug and receptacle where individual overload protection is omitted, the rating of the attachment plug and receptacle shall not exceed _____ amperes at 250 volts.

   a. 15          b. 20                    c. 25                    d. 30

4. An overcurrent device rated at not more than _____ percent of the rated primary current shall be permitted in the primary circuit if a control circuit transformer rated primary current is less than 2 amperes.

   a. 175          b. 300                    c. 500                    d. 800

5. A disconnecting means shall plainly indicate whether it is in the open or closed position.

   a. True          b. False

6. Current in the common conductor of a 2-phase, 3-wire system will be _____ times the values given in Table 430.249.

   a. 2          b. 1.732                    c. 1.25                    d. 1.41

7. The ampacity of conductors supplying a motor rated for intermittent duty with a 15-minute duty cycle must be no less than _____ percent of the motor nameplate current.

   a. 110          b. 90                    c. 150                    d. 85

8.  Providing other conditions are met, a group of motors rated for 1 horsepower or less are permitted on a 20A, 120-volt branch circuit if the full-load rating of each motor does not exceed _____ amperes.

    a. 3              b. 5              c. 6              d. 7

9.  The number of overload units required for a single-phase motor being fed from a 3-phase system is _____ in an ungrounded conductor.

    a. two            b. one            c. three          d. not required

10. A _____ winding start induction motor is arranged so that one-half of its primary winding can be energized initially, and, subsequently, the remaining half can be energized, where both halves then carry equal amounts of current.

    a. standard       b. premium        c. standard part  d. dual

11. A controller must be capable of starting and stopping the motor it controls and shall be capable of interrupting the locked-rotor current of the motor.

    a. True           b. False

12. A set of 18' feeder tap conductors is required to have an ampacity at least one-fourth that of the feeder conductors.

    a. True           b. False

13. Where automatic restarting of a motor can result in personal injury, a motor overload device that can restart a motor automatically after overload tripping shall not be installed.

    a. True           b. False

14. A No. 10 copper control circuit conductor extending beyond a motor control enclosure may be protected by the motor branch-circuit protective device when the rating of the device does not exceed _____ amperes.

    a. 160            b. 75             c. 140            d. 90

15. The disconnecting means for a torque motor is permitted to be a circuit breaker.

    a. True           b. False

16. When a synchronous motor has a .90 power factor, the full-load current values listed in Table 430.250 are required to be increased by _____ percent.

    a. 110            b. 125            c. 140            d. 150

17. Torque motors are rated for operation at standstill.

    a. True           b. False

18. Where designed for such starting, an adjustable-speed motor is permitted to be started under a weakened field.

    a. True              b. False

19. Where the opening of a control circuit would create a hazard, overcurrent protection for a control circuit transformer shall be omitted.

    a. True              b. False

20. Where dual element fuses are used for the overcurrent protection of both windings of a part-winding motor, the rating of the fuses shall not exceed _____ percent of the motor full-load current.

    a. 125          b. 150              c. 175              d. 200

21. Open motors that have commutators or collector rings are permitted to be installed on wooden floors.

    a. True              b. False

22. Additional overload protection shall not be required when the power conversion equipment is marked to indicate that motor overload protection is included.

    a. True              b. False

23. Where dual element fuses are used to protect a wound rotor motor, the rating of the fuse must not exceed _____ percent of the motor's full-load current.

    a. 300          b. 175              c. 250              d. 150

24. The minimum wire-bending space for four parallel No. 300 kcmil conductors (per terminal) leaving the terminals of an enclosed motor controller opposite the controller's terminals is _____ inches.

    a. 19           b. 12              c. 14              d. 17

25. The disconnecting means for motor circuits rated less than 600 volts shall have an ampere rating of at least _____ percent of the full-load current rating of the motor.

    a. 150          b. 225              c. 115              d. 300

26. The nameplate current for high torque motors must be used instead of the full-load current values listed in Tables 430.247–250.

    a. True              b. False

27. For stationary motors rated for 2 hp or less, a general snap switch that is exclusively used for AC circuits can be used as a controller when the motor full-load current rating is not more than _____ of the ampere rating of the switch.

   a. 25 percent          b. 60 percent          c. 80 percent          d. 110 percent

28. The circuit of a control apparatus or system that carries the electric signals directing the performance of the controller but that does not carry the main power current is known as a(n) _____.

   a. control circuit          b. remote control circuit          c. motor control circuit
   d. auxiliary control circuit

29. In accordance with Article 430, a controller is any switch or device that is normally used to start and stop a motor by _____ the motor circuit current.

   a. breaking          b. making          c. either a or b          d. both a and b

30. The values given in Tables 430.247–250, including notes, shall be used to determine the ampacity of conductors, the ampere rating of switches and overcurrent protection of motors instead of the actual _____ rating marked on the motor.

   a. horsepower          b. current          c. locked-rotor          d. nameplate

31. The rating of a motor is based on its rated _____.

   a. horsepower          b. full-load current          c. locked-rotor current          d. torque

32. Overload in an electrical apparatus is an operating overcurrent that, when it persists for a sufficient length, would cause damage or dangerous overheating of the apparatus.

   a. True          b. False

33. A metal junction box that is likely to become energized is permitted to be separated from a motor with selected raceways enclosing conductors with stranded leads not larger than No. _____ AWG.

   a. 14          b. 8          c. 10          d. 12

34. A controller shall be permitted to be an attachment plug and receptacle for a portable motor not exceeding _____ hp.

   a. 1/3          b. 1/4          c. 1/2          d. 3/4

35. An inverse time circuit breaker is permitted to be operated both manually and automatically.

   a. True          b. False

36. Control circuit devices with screw-type pressure terminals used with a No. 14 AWG or smaller copper conductor shall be torqued to a minimum of _____ pound-inches unless identified for a different torque.

    a. 10                b. 7                     c. 9 3/4                     d. 6.5

37. When a motor control apparatus exceeds 600 volts, the ultimate trip current of overload relays shall not exceed _____ of the controller's continuous current rating.

    a. 115               b. 125                 c. 150                 d. 175

38. A fuse is required in each ungrounded conductor and also in the grounded conductor if the supply system is _____ with one conductor grounded.

    a. 2-wire, single phase

    b. 3-wire, single phase

    c. 3-wire, three phase

    d. 4-wire, three phase

39. When a resistor is used for light intermittent duty, the ampacity of the conductors between the resistor and controller shall not be less than _____ percent of the full-load secondary current.

    a. 85               b. 75                  c. 65                  d. 55

40. Separate motor overload protection shall be based on a motor's in-rush current rating.

    a. True               b. False

41. Branch circuit conductors that supply a single motor used in a continuous duty application shall have an ampacity not less than 125 percent of the _____ rating.

    a. motor's kVA per horsepower            b. motor's nameplate current

    c. motor's full-load current                d. motor's horsepower

42. Overload protection for motors is increased to 125 percent of the motor's nameplate current when the motor is marked with _____.

    I.  a service factor that is less than or equal to 1.15

    II.  a temperature rise of 40°C or less

    III.  a service factor that is equal to or greater than 1.15

    IV.  a temperature rise of 40°C or greater

    a. I and II             b. II and III            c. III and IV           d. I and IV

43. Instantaneous-trip circuit breakers may include a damping means to accommodate a transient motor inrush current without nuisance tripping of the circuit breaker.

    a. True               b. False

44. The maximum full-load current for a 9-lead, 75-hp, 3-phase motor rated for 460 volts is _____ amperes.

    a. 168        b. 44          c. 71          d. 110

45. Types FMC or LFMC not exceeding _____ in length shall be permitted to be employed for raceway connection to a motor terminal enclosure.

    a. 6'         b. 3'          c. 7'          d. 4'

46. Other than instantaneous-trip circuit breakers and molded case switches, a controller shall have horsepower ratings at the application voltage not lower than the horsepower rating of the motor.

    a. True               b. False

47. If fuses or inverse time circuit breakers rated or set at not over _____ percent of the full-load current of the motor are located in the circuit so as to be operative during the starting period of the motor, overload protection for a nonautomatically started motor shall be permitted to be shunted or cut out of the circuit during the starting period of the motor.

    a. 150        b. 200         c. 325         d. 400

48. Any motor application shall be considered as continuous duty unless the nature of the apparatus it drives is such that the motor will not operate continuously with load under any condition of use.

    a. True               b. False

49. The frames of portable motors that operate at over 150 volts to ground shall be guarded or grounded.

    a. True               b. False

50. Motors of _____ are not required to be marked with their horsepower rating.

    a. variable torque drive systems            b. arc welders
    c. air conditioning and refrigeration equipment        d. generators

51. The selection of conductors between a controller and a part-winding motor shall be based on _____ percent of the motor's full-load current.

    a. 50         b. 80          c. 125         d. 150

52. The ultimate trip current of a thermally protected motor shall not exceed _____ percent when the motor full-load current is greater than 20 amperes.

    a. 115              b. 140                    c. 156                    d. 170

53. In accordance with NEC 240.4(B), torque motor branch circuits shall be protected at the motor's _____ current rating.

    a. full-load        b. nameplate              c. locked-rotor           d. overload

54. An autotransformer starter shall provide a(n) _____ position.

    a. running

    b. off

    c. minimum of one starting position

    d. all of the preceding

55. Conductors connecting a motor controller to separately mounted power accelerating and dynamic braking resistors in the armature circuit shall have an ampacity not less than 75 percent when the duty cycle is _____ on, _____ off.

    a. 15, 15           b. 10, 70                 c. 15, 30                 d. 15, 45

56. Overtemperature can occur if a motor that uses external forced air or liquid cooling systems fails.

    a. True             b. False

57. A motor feeder supplying specific fixed motor loads consisting of conductor sizes based on NEC 430.24 requires an overcurrent protective device having a rating greater than the largest rating of the branch-circuit short-circuit and ground-fault protective device of any motor supplied by the feeder plus the sum of the remaining motor's full-load currents.

    a. True             b. False

58. When supply conductors are isolated from the busbars of a motor control center by a barrier, they are permitted to travel _____ through vertical sections.

    a. vertically       b. from front to rear     c. horizontally           d. uninsulated

59. If the rating of time-delay fuses specified in Table 430.52 is not sufficient to start a motor, the rating of the fuses can be increased up to _____.

    a. 225 percent      b. 250 percent            c. 300 percent            d. 400 percent

60. The kVA per horsepower with locked rotor ranges from _____ for motors with code letter E.

a. 6.3 to 7.09     b. 3.1 to 3.54     c. 4.5 to 4.99     d. 5 to 5.59

61. The conductors between a stationary motor rated 1 hp or less and a separate terminal enclosure are not permitted to be smaller than a No. _____ AWG.

a. 14     b. 16     c. 12     d. 18

62. Where overload protection is not sufficient to start a motor or to carry a load, higher size sensing elements are permitted.

a. True     b. False

# ● ARTICLE 440–AIR-CONDITIONING AND REFRIGERATING EQUIPMENT

1. The rating of attachment plugs and receptacles for motor-compressors shall not exceed 20 amperes at 250 volts.

   a. True                 b. False

2. A disconnecting means serving a hermetic refrigerant motor-compressor shall be selected on the basis of _____.

   a. the branch circuit selection current

   b. the nameplate-rated load current

   c. 175 percent of the motor-compressor rated-load current

   d. 225 percent of the branch-circuit selection current

   e. the larger of a or b

   f. d if c is insufficient

3. Where not an integral part of an attachment plug, LCDI or AFCI protection must be located in a room air conditioner's power supply cord within _____ of the attachment plug.

   a. 6"        b. 18"        c. 12"        d. 24"

4. The locked-rotor current of each _____ shall be marked on the motor-compressor nameplate.

   a. single-phase motor-compressor having a rated-load current more than 4.5 amperes at 230 volts

   b. each two-phase motor

   c. single-phase motor-compressor having a rated-load current more than 9 amperes at 115 volts

   d. each three-phase motor

   e. only a and c

   f. only b and c

   g. a, b, c and d

   h. a or c and b and d

5. Conductors supplying a group of motor-compressors with or without an additional load(s) shall have an ampacity not less than the sum of the rated-load or branch-circuit selection current ratings, whichever is larger, of all the motor-compressors, plus the full-load currents of the other motors, plus _____ percent of the highest motor or motor-compressor rating in the group.

   a. 25        b. 50        c. 110        d. 125

6.  The rated load current for a hermetic refrigerant motor-compressor is the current resulting when the motor-compressor is operated at the rated _____ of the equipment it serves.

    a. load          b. voltage          c. frequency          d. a and b          e. a and c
    f. b and c          g. a, b, or c          h. a, b, and c

7.  The branch circuit overcurrent protection for a motor-compressor shall not exceed _____ percent of the motor-compressor rated-load current or branch circuit selection current, whichever is greater.

    a. 125          b. 150          c. 175          d. 225

8.  Where so marked, the rated-load current shall be used instead of the branch circuit selection current in determining the highest rated motor-compressor.

    a. True          b. False

9.  A flexible supply cord used for room air conditioners shall not exceed _____ feet for units rated for 120 volts and _____ feet for units rated for 208 or 240 volts.

    a. 6, 10

    b. 4, 8

    c. 12, 7

    d. none of the preceding

10. The value of branch circuit selection current will always be _____ the marked rated-load current.

    a. equal to          b. less than          c. greater than          d. a or b          e. a or c

11. Where lighting units or other appliances are used, a cord and plug connected room air-conditioner shall not exceed _____ of the rating of the branch circuit.

    a. 80          b. 50          c. 60          d. 25

12. An overload device selected to trip up to _____ percent of a motor-compressor rated load current can be used to protect against overload.

    a. 115          b. 125          c. 130          d. 140

13. As applicable, the rules of Article(s) _____ shall apply to air-conditioning and refrigerating equipment that does not incorporate a hermetic refrigerant motor-compressor.

    a. 422          b. 424          c. 430          d. a and b
    e. a and c          f. b and c          g. a, b or c          h. a, b, and c

# ● ARTICLE 445–GENERATORS

1. Generators operating at _____ volts or less and driven by individual motors shall be considered as protected by the overcurrent device protecting the motor if these devices will operate when the generators are delivering not more than 150 percent of their full-load rated current.

   a. 150                 b. 300                 c. 65                 d. 600

2. Only if the overcurrent device actuated by the entire current generated, other than the current in the shunt field, is a two-wire DC generator permitted to have overcurrent protection in one conductor.

   a. True                 b. False

3. The ampacity of the conductors from the generator terminals to the first distribution devices containing overcurrent protection shall not be less than _____ of the nameplate current rating of the generator.

   a. 100                 b. 150                 c. 115                 d. 125

# ● ARTICLE 450–TRANSFORMERS AND TRANSFORMER VAULTS (INCLUDING SECONDARY TIES)

1. A transformer installed indoors and rated for over 35,000 volts shall be furnished with a _____ .

   a. pressure-relief vent

   b. liquid confinement area

   c. both a and b

   d. chimney or flue

2. Dry-type transformers shall be provided with a _____ case that provides protection against the accidental insertion of foreign objects.

   a. totally enclosed                    b. noncombustible moisture-resistant
   c. ventilated moisture retardant       d. secured

3. A vault enclosing a 150kVA transformer must be provided with _____ .

   a. ventilation        b. a drain        c. fire-resistant walls        d. sprinklers

4. The phase current in a grounding autotransformer is 2/3 the neutral current.

   a. True               b. False

5. The floors of vaults in contact with the earth shall be of concrete that is not less than 6 inches thick.

   a. True               b. False

6. All ventilation openings to the indoors shall be provided with automatic closing fire dampers with a standard fire rating of not less than 1 hour.

   a. True               b. False

7. Transformers exceeding 112-1/2 kVA shall not be located within 12 inches of combustible materials of buildings unless the transformer has _____ insulation systems or higher.

   a. Design B           b. Class 155           c. NEMA 12           d. 150°C

8. Nonlinear loads can increase heat in a transformer without operating its overcurrent protective device.

   a. True               b. False

9. The term "fire resistant" means a construction having a minimum fire rating of _____.

    a. 30 minutes          b. 1 hour                c. 1.5 hours             d. 2 hours

10. When transformer vaults are protected with automatic sprinklers or carbon dioxide, the minimum fire rating of a door can be reduced to _____ hour(s).

    a. 2                   b. 1.5                   c. .75                   d. 1

11. Where secondary ties are used, an overcurrent device rated or set at not more than 150 percent of the rated secondary current of the transformer shall be provided in the secondary connections of each transformer.

    a. True               b. False

12. The word transformer shall mean _____ that is identified by a single nameplate.

    a. a single transformer

    b. a polyphase bank of two single-phase transformers operating as a unit

    c. a polyphase bank of three single-phase transformers operating as a unit

    d. all of the preceding

13. The roofs and walls of a transformer vault must have a minimum fire resistance of _____ hour(s).

    a. 1.5                   b. 3                     c. 1                    d. 2

14. An overcurrent device shall be rated or set at a current not exceeding 125 percent of the autotransformer's continuous per-phase current rating or _____ percent of any series-connected devices in the autotransformer's neutral connection.

    a. 42                   b. 80                   c. 125                d. 150

15. Where rated for over _____, a dry-type transformer shall be installed in a vault.

    a. 112kVA              b. 10 percent impedance     c. 300A              d. 35kV

16. A secondary tie is a circuit operating at _____ or less, between phases that connect two power sources or power supply points.

    a. 300                 b. 600                 c. 1000              d. 200

17. In a non-maintenance area, the rating of a circuit breaker used to protect the primary side of a 2.4kV transformer with a 7 percent impedance must not exceed _____ percent of the transformer's rated current.

    a. 225                 b. 250                 c. 300              d. 400

### ● ARTICLE 455–PHASE CONVERTERS

1. When a phase converter supplies either fixed or variable loads, the overcurrent protection shall not exceed 125 percent of the phase converter nameplate single-phase input amperes.

   a. True            b. False

2. Capacitors that are not an integral part of a rotary-phase conversion system but that are installed for a motor load shall be connected to the supply side of the motor overload protective device.

   a. True            b. False

3. When a phase converter supplies specific fixed loads, and the single-phase supply conductor's ampacity is less than 125 percent of the phase converter nameplate single-phase input full-load amperes, the conductors shall have an ampacity not less than _____ of the sum of the full-load, 3-phase current rating of the motors and other loads served.

   a. 125 percent      b. 175 percent      c. 250 percent      d. 300 percent

4. A device without rotating parts, sized for a given 3-phase load to permit operating from a single-phase supply, is called a _____.

   a. stationary phase converter          b. static phase converter
   c. fixed phase converter          d. nonrotary phase converter

5. Where the input and output voltages of a phase converter differ, the current shall be _____ the ratio of the output to input voltage.

   a. divided into      b. multiplied by      c. added to      d. subtracted from

6. Disconnecting means for all ungrounded supply conductors to a phase converter must be a _____.

   a. horsepower-rated switch

   b. circuit breaker

   c. molded-case switch

   d. all of the preceding

7. The ampacity of phase converters used to supply _____ loads must not be less than 125 percent of the phase converter nameplate single-phase input full-load amperes.

   a. intermittent      b. variable      c. fixed      d. continuous

# ARTICLE 460–CAPACITORS

1.  The rating of an overcurrent device protecting a capacitor bank must be as low as practicable.

    a. True     b. False

2.  The residual voltage of a capacitor shall be reduced to 50 volts or less within _____ after the capacitor is disconnected from the source of supply.

    a. 30 seconds  b. 45 seconds  c. 1 minute  d. 1.5 minutes

3.  The ampacity of capacitor circuit conductors shall not be less than _____ percent of the rated current of the capacitor.

    a. 115    b. 125    c. 135    d. 145

4.  Capacitors containing more than _____ gallons of flammable liquid shall be enclosed in vaults or outdoor fenced enclosures.

    a. 3     b. 4     c. 5     d. 6

5.  A means shall be provided to reduce the residual voltage of a capacitor rated for over 600 volts to _____ volts or less within 5 minutes after the capacitor is disconnected from the source of supply.

    a. 30     b. 50     c. 75     d. 100

6.  When a motor installation includes a capacitor connected on the load side of the motor overload device, the rating or setting of the motor overload device shall be based on the _____ of the motor circuit.

    a. ampacity

    b. rated current

    c. improved power factor

    d. supplemental protection

# ARTICLE 470–RESISTORS AND REACTORS

1. A _____-inch clearance must be maintained between resistors and reactors and combustible materials.

   a. 3                b. 6                c. 9                d. 12

2. Insulated conductors used for connections between resistance elements and controllers shall be suitable for an operating temperature of not less than _____.

   a. 60°C             b. 75°F            c. 90°C             d. 150°F

3. Resistors and reactors shall not be placed where exposed to physical damage.

   a. True             b. False

# ● ARTICLE 480–STORAGE BATTERIES

1. Overcurrent protection shall not be required for conductors from a battery rated less than 50 volts if the battery provides power for _____.

   a. ignition      b. starting      c. control of prime movers      d. all choices apply

2. As an alternative, sealed cells shall be equipped with a _____ to prevent excessive accumulation of gas pressure.

   a. flame arrestor      b. alarm      c. vent hole      d. pressure relief vent

3. The nominal voltage of a battery is the voltage computed on the basis of 2.5 volts per cell for the lead-acid type and 2.1 volts per cell for alkali type.

   a. True      b. False

4. A battery comprised of one or more rechargeable cells of the lead-acid, nickel-cadmium, or other rechargeable electrochemical types is defined as a(n) _____ battery.

   a. rechargeable      b. alkali      c. storage      d. direct current

5. No additional insulation support is required for cells in rubber or composition containers where the nominal voltage of all cells in series does not exceed _____ volts.

   a. 24      b. 60      c. 130      d. correct choice not provided

6. Cells in jars of conductive material shall be installed in trays of nonconductive material with not more than 10 cells in the series circuit in any one tray.

   a. True      b. False

# ● ARTICLE 490–EQUIPMENT, OVER 600 VOLTS, NOMINAL

1.  Where located outdoors, the minimum clearance for live phase-to-phase circuits rated for 23kV is _____.

    a. 9"              b. 12"              c. 15"              d. 18"

2.  Circuit breakers rated for over 600 volts, when installed indoors, shall be _____

    a. mounted in metal-enclosed units

    b. permitted to be open mounted where accessible only to qualified persons

    c. mounted in fire-resistant cell-mounted units

    d. all of the preceding

3.  Where fuses rated for over 600 volts can be energized by backfeed, a sign shall be placed on the enclosure door identifying this hazard.

    a. True              b. False

4.  An electrode-type boiler branch circuit shall be provided with means for the detection of the sum of the neutral conductor and equipment grounding conductor currents and shall trip the circuit-interrupting device if the sum of those currents exceeds an instantaneous value of _____ percent of the boiler full-load current.

    a. 7.5              b. 12.5              c. 20              d. 25

5.  Oil-filled cutouts shall be located so that they are readily and safely accessible for re-fusing, with the top of the cutout not over _____ feet above the floor or platform.

    a. 3              b. 4              c. 5              d. 6

6.  A 2.4kV electrode-type boiler can only be supplied from a _____.

    a. 3-phase, 3-wire, corner-grounded delta system

    b. 3-phase, 4-wire wye system

    c. 3-phase, 4-wire, solidly grounded wye system

    d. 3-phase, 4-wire delta system

7.  Other than transformers, the installation of electrical equipment containing more than _____ gallons of flammable oil per unit shall meet the requirements of Parts II and III of Article 450.

    a. 3              b. 5              c. 8.5              d. 10

8.  Control and instrument transfer switch handles affiliated with high voltage equipment shall be located no higher than _____ in either the open or closed position when the handles require more than 50 lbs of force.

    a. 66"             b. 72"             c. 69"             d. 78"

9.  Power fuses of the vented type can be used indoors.

    a. True            b. False

10. Proper switching sequence for regulators shall be ensured by _____.

    I. mechanical interlocks

    II. mechanically sequenced regulator bypass switch(es)

    III. prominently displayed switching procedures

    a. I and II        b. II and III      c. I and III       d. either I, II, or III

11. Where more than one switch is permitted as the disconnecting means for one set of fuses, a _____ shall be placed at the fuses, identifying the presence of more than one source.

    a. diagram

    b. warning label

    c. conspicuous sign

    d. all choices apply

# CHAPTER 5–SPECIAL OCCUPANCIES

## ⊙ ARTICLE 500–HAZARDOUS (CLASSIFIED) LOCATIONS, CLASSES I, II, AND III, DIVISIONS 1 AND 2

1.  Flammable gas, flammable liquid–produced vapor, or combustible liquid-produced vapor mixed with air that may burn or explode, having either a MESG value greater than 0.75 mm or a MIC ratio greater than 0.80 are classified under Group _____.
    a. A      b. D      c. F      d. C      e. none of the choices

2.  Where ignitable concentrations of flammable gases or flammable liquid-produced vapor may frequently exist because of repairs, maintenance operation, or leakage, an area must be classified as _____.
    a. Cl III, Div 2      b. Cl II, Div 1      c. Cl I, Div 1      d. Cl III, Div 2

3.  All threaded conduit or fittings referred to in a classified area must be threaded with a cutting die that provides a _____ taper per foot.
    a. 1"      b. 16"      c. 3/4"      d. 1/2"      e. 1 3/4"

4.  A dusttight enclosure is constructed so that dust will not enter under unspecified test conditions.
    a. True      b. False

5.  Equipment in a classified area shall be marked to specify the temperature class or operating temperature at _____ ambient temperature.
    a. 167°F      b. 60°C      c. 40°F      d. 40°C

6.  Electrical equipment suitable for Class I, Division 2 locations is not permitted in a Class I, Division 1 location that is so classified due to inadequate ventilation.
    a. True      b. False

7.  A dust-ignitionproof enclosure excludes dusts and does not permit arcs, sparks, or heat.
    a. True      b. False

8.  Atmospheres containing combustible metal dusts, including aluminum, magnesium, and their commercial alloys, are classified under Group _____.
    a. A      b. D      c. F      d. C      e. none of the preceding

9. Carbon disulfide is one chemical that requires safeguards beyond those required for any of the Class I groups because of its _____.

    I. high ignition temperature

    II. small joint clearance

    III. low ignition temperature

    a. I and II          b. II and III          c. I and III

10. Equipment that is sealed by means of fusion such as soldering, brazing, welding, or the fusion of glass to metal is called _____.

    a. permanently sealed       b. tightly sealed       c. hermetically sealed

# ● ARTICLE 501–CLASS I LOCATIONS

1. When metal conduit pass completely through a Class I, Division 1 location and contain no fittings, couplings, unions, or boxes, a conduit seal is not required for fittings that are less than 12" beyond each boundary if the termination points of the unbroken conduit are located in an unclassified area.

   a. True                  b. False

2. Flexible cord listed for extra-hard usage and provided with listed bushed fittings can be used when making flexible connections.

   a. True                  b. False

3. Receptacles and attachment plugs shall be of the type providing for connection to the equipment grounding conductors of a flexible cord.

   a. True                  b. False

4. Motors, generators, and other rotating electric machinery must be which of the following types?

   a. totally enclosed     b. totally enclosed inert gas-filled     c. flammable liquid submersible
   d. all choices apply

5. Splices and taps are not allowed to be made in fittings intended only for sealing with compound.

   a. True                  b. False

6. Process control instruments are permitted to be connected through a flexible cord, attachment plug, and receptacle, providing the current does not exceed _____.

   a. 4A @ 50V          b. 6A @ 110V                c. 3A @ 120V               d. 5A @ 240V

7. Only listed _____ are permitted between conduit seals and the point where conduit leaves Division 2 locations.

   a. couplings          b. reducers                c. fittings               d. boxes

8. Sealing compound shall not have a melting point less than 93°C (200°F) in _____.

   a. Class I, Division 1 locations only          b. Class I, Division 2 locations only
   c. Class I, Divisions 1 and 2 locations only   d. Both a and b

9. Threaded _____ are approved wiring methods in Class I, Division 2 locations.

   a. Rigid metal conduit      b. IMC      c. both a and b      d. neither a nor b

10. Transformers and capacitors not containing a liquid that will burn shall be installed in a vault.

    a. True      b. False

11. Seals that are installed as close as possible to an enclosure will reduce problems with _____ the dead airspace in pressurized conduit.

    a. vaporizing      b. sealing      c. purging      d. pressuring

12. Seals are provided in conduit and cable systems to minimize the passage of _____ and _____ and prevent the passage of flames from one portion of the electrical installation to another through the conduit.

    a. corrosion, oxidation      b. vapors, gases      c. air, fumes      d. arcs, electrolysis

13. Multiwire branch circuits are permitted in Class I, Division 2 locations.

    a. True      b. False

14. Flexible connections are not permitted in Class I locations.

    a. True      b. False

15. The cross-sectional area of the conductors permitted in a seal must not exceed _____

    a. 60 percent

    b. 53 percent

    c. 40 percent

    d. 31 percent

    e. 25 percent

    of the cross-sectional area of a(n) _____ conduit of the same trade size unless it is specifically identified for a higher percentage of fill.

    a. nonmetallic

    b. intermediate metal

    c. rigid metal

    d. flex metal specifically identified for a higher percentage fill

16. Bonding jumpers with proper fittings or other approved means of bonding shall apply to all intervening raceways, fittings, boxes, and enclosures between _____ locations and the point of grounding for service equipment or of a separately derived system.

    a. Class I only

    b. Class II only

    c. Both Class I and II

    d. neither Class I nor II

17. Enclosures for windings of impedance coils, solenoids, or transformers are permitted to be of the _____ type.

    a. general-purpose    b. explosionproof    c. current-limiting    d. sealtight

18. Factory-sealed enclosures are permitted to serve as a seal for another adjacent explosion-proof enclosure that is required to have a conduit seal.

    a. True    b. False

19. When two or more explosionproof enclosures requiring conduit seals are connected by nipples or runs of conduit not more than _____, a single conduit seal in each nipple connection or run of conduit shall be considered sufficient if located not more than _____ from either enclosure.

    a. 24", 24"    b. 36", 18"    c. 24", 18"    d. 18", 36"

20. Enclosures for instruments must be _____.

    a. explosionproof    b. purged    c. pressurized    d. a and c    e. b and a
    f. c and b    g. none of the preceding    h. all of the preceding

21. In Class I, Division 1 and 2 locations, the minimum thickness of sealing compound in a completed seal shall not be less than trade size of the sealing fitting and, in no case, less than _____.

    a. 1/2"    b. 5/8"    c. 3/4"    d. 1"

22. Conduit seals are required to be installed within _____ from an enclosure.

    a. 24"    b. 18"    c. 30"    d. 12"

23. In Class I, Division 2 locations, segments of aboveground conduit systems are not required under certain conditions to be sealed if passing into an unclassified location.

    a. True    b. False

24. Pendant luminaires shall be suspended by and supplied through _____ or _____ stems.

   a. EMT, threaded rigid metal conduit

   b. threaded steel intermediate conduit, PVC

   c. threaded rigid metal conduit, threaded steel intermediate conduit

   d. threaded steel intermediate conduit, EMT

25. When entering an explosionproof enclosure, conduit entries must be at least _____ in trade size.

   a. 1"                b. 1 1/2"                c. 2"                d. 2 1/2"

26. Where heaters are used for electrically heated utilization equipment, the heaters shall be identified for which classified location?

   a. Class II, Division 2          b. Class I, Division 1          c. Class I, Division 2
   d. Class II, Division 1

27. Surge protective devices shall be of a type designed for specific duty.

   a. True                b. False

28. In motors and generators, the exposed surface of space heaters used to prevent condensation of moisture during shutdown periods shall not exceed _____ percent of the ignition temperature in degrees Celsius.

   a. 125                b. 80                c. 75                d. 45

29. Cables with an unbroken _____ shall be permitted to pass through a Class I, Division 2 location without seals.

   a. air/vaportight continuous sheath          b. gas/airtight metallic sheath
   c. gas/vaportight nonmetallic sheath          d. gas/vaportight continuous sheath

30. Only _____ rigid metal conduit is permitted as an acceptable wiring method.

   a. tradesize 2                b. threaded                c. PVC coated                d. buried

31. Multiconductor cable in conduit must be considered as a single conductor if the cable is incapable of transmitting gases or vapors through the cable core.

   a. True                b. False

# ◉ ARTICLE 502–CLASS II LOCATIONS

1. A _____ enclosure shall be provided with switching mechanisms associated with control transformers, solenoids, impedance coils, and resistors.

    a. sealtight          b. raintight               c. airtight          d. dusttight

2. The seams and joints of metal pipes that are used for ventilation must be _____.

    a. screwed and bolted     b. brazed and soldered     c. riveted and welded
    d. sealed and wrapped     e. none of the preceding

3. Where flexible connections are required, which of the following wiring methods can be used?

    I. liquidtight flexible nonmetallic conduit with listed fittings

    II. dusttight flexible connectors

    III. liquidtight flexible metallic conduit with listed fittings

    IV. flexible cord listed for extra-hard usage with bushed fittings

    a. Only I and III     b. only II          c. only II and IV          d. only I–III
    e. all of the preceding methods

4. As an approved wiring method, fittings and boxes must be provided with threaded bosses for connection to conduit or cable terminations and shall be _____.

    a. airtight          b. compound-sealed     c. dusttight          d. waterproof

5. Liquidtight flexible metal conduit shall not be used as the sole ground-fault current path.

    a. True               b. False

6. Which enclosure is not permitted for motors, generators, and other rotating electrical machinery?

    a. totally enclosed nonventilated          b. totally enclosed air-cooled
    c. totally enclosed fan-cooled             d. totally enclosed pipe-ventilated

7. Transformers or capacitors shall be installed in a location where _____ from magnesium, aluminum, aluminum bronze powders, or other metals of similarly hazardous characteristics may be present.

    a. residue           b. dust             c. film             d. deposits

8.  An 11' horizontal raceway is permitted to prevent the entrance of dust into a dust-ignitionproof enclosure when raceway provides communication between an enclosure that is required to be dust-ignitionproof and one that is not.

    a. True                  b. False


9.  Which of the following wiring method is not allowed in Division 2 locations?

    a. EMT            b. IMC            c. RMC            d. general purpose wireway

# ⦿ ARTICLE 503–CLASS III LOCATIONS

1.  Luminaires that are permanently installed are required to be marked to show the maximum allowable wattage of lamps without exceeding an exposed surface temperature of 165°C under normal conditions of use.

    a. True                b. False

2.  Storage battery charging equipment shall be located in separate rooms built or lined with substantial _____ materials.

    a. sealtight       b. noncombustible       c. fireproof       d. self-ventilating

3.  In Class III locations, the maximum surface temperatures under operating conditions shall not exceed _____ for equipment that is not subject to overloading.

    a. 165°F       b. 120°C       c. 329°F       d. 248°C

4.  Contact conductors shall be protected against accidental contact with foreign objects and shall be located or guarded so as to be inaccessible to unauthorized persons.

    a. True                b. False

5.  Which of the following wiring methods are not allowed in Division 2 locations?

    I. dusttight auxiliary gutters

    II. IMC

    III. EMT

    IV. Type BX cable with listed termination fittings

    V. Rigid metal conduit

    a. I and V       b. II and IV       c. II and III       d. I and IV       e. III and V

# ⦿ ARTICLE 504–INTRINSICALLY SAFE SYSTEMS

1. Conductors of intrinsically safe circuits shall not be placed in any raceway, cable tray, or cable with conductors of any nonintrinsically safe circuit.

   a. True            b. False

2. Intrinsically safe conductors are permitted to be identified by _____ colors.

   a. yellow       b. orange       c. light-blue       d. red

3. Conductors and cables supplying intrinsically safe circuits not in raceways or cable trays shall be separated by at least _____ and secured from conductors and cables of any nonintrinsically safe circuits.

   a. 6"       b. 2"       c. 8"       d. 4"

4. A circuit in which any spark or thermal effect is incapable of causing ignition of a mixture of flammable or combustible material in air under prescribed test conditions is referred to as a(n) _____.

   a. different intrinsically safe circuit       b. intrinsically safe circuit
   c. ignition free circuit       d. nonintrinsically safe circuit

5. Where separated by enclosures, walls, partitions, or floors, intrinsic safety circuit labels must be spaced no more than _____ feet apart.

   a. 10       b. 18       c. 25       d. 30

6. Intrinsically safe circuits shall be identified at _____ and _____ locations.

   a. spliced, terminating       b. junction, terminal       c. testing, servicing
   d. maintenance, repair

7. All intrinsically safe apparatus and associated apparatus shall be _____.

   a. labeled       b. marked       c. listed       d. approved

# ● ARTICLE 505–CLASS I, ZONE 0, 1, AND 2 LOCATIONS

1. For equipment with metric threaded entries, such entries shall be _____ as being metric, or _____ adapters to permit connection to conduit or NPT-threaded fittings shall be provided with the equipment.

   a. labeled, approved

   b. recognized, provided with

   c. identified, listed

   d. listed, allowed

2. MESG stands for _____.
   a. minimum electrical service ground     b. maximum experiment safe gap
   c. medium explosion safety guard     d. maintenance exclusive separate guide

3. Provided the termination is by an approved means to minimize the entrance of gases or vapor and prevent propagation of flame into the cable core, shielded cables and twisted pair cables shall not require the removal of the shielding material or separation of the twisted pairs.

   a. True       b. False

4. Rooms and areas containing ammonia refrigeration systems that are equipped with adequate mechanical ventilation may be classified as "classified" locations.

   a. True       b. False

5. Equipment that is listed for a Zone 1 location is only permitted in a Zone 1 location of the same gas or vapor.

   a. True       b. False

6. The type of protection in which the enclosure will withstand an internal explosion of a flammable mixture that has penetrated into the interior (without suffering damage and without causing ignition) through any joints or structural openings in the enclosure of an external explosive gas atmosphere consisting of one or more of the gases or vapors for which it is designed is recognized as _____.
   a. fireproof "c"     b. flameproof "d"     c. explosionproof "a"     d. ignitionproof "b"

7. For protection type "e," field wiring conductors must be _____.
   a. insulated     b. covered       c. grounded       d. minimized
   e. none of the choices apply

8. Locations where ignitable concentrations of flammable gases or vapors may frequently exist because of repairs, maintenance, or leakage are classified as _____.

    a. Class I, Zone 2      b. Class I, Zone 1      c. Class I, Zone 0      d. Class I, Zone 3

9. Gas or vapor leakage and propagation of flames may occur through the interstices between the standard stranded conductors larger than _____.

    a. No. 12 AWG      b. No. 10 AWG      c. No. 8 AWG      d. No. 6 AWG
    e. none of the choices apply

10. A protection technique utilizing stationary gas detectors in industrial establishments is called a _____.

    a. gas detection system                      b. stationary gas detection system
    c. combustible gas detection system          d. utilization protection system

11. Electrical metallic tubing is an approved wiring method in Class I, Zone 2 locations.

    a. True                b. False

12. The symbol AEx identifies _____.

    a. equipment bought with added and extra protection

    b. advanced exterior enclosures

    c. equipment built to American standards

    d. areas under extreme precaution

13. The type of protection where electrical equipment, in normal operation, is not capable of igniting a surrounding explosive gas atmosphere and a fault capable of causing ignition is not likely to occur provides which of the following types of protection?

    a. "o"                b. "m"                c. "q"                d. "n"

14. The minimum distance of obstruction from flameproof "d" flange openings under Gas Group IIB is _____.

    a. 1 1/2"             b. 1 37/64"           c. 1 3/16"            d. 25/64"

# ● ARTICLE 506–ZONE 20, 21, AND 22 LOCATIONS FOR COMBUSTIBLE DUSTS OR IGNITIBLE FIBERS/ FLYINGS

1. Locations that normally are classified as Zone 21 can fall into Zone 22 when measures are employed to prevent the formation of explosive _____ mixtures.

   a. hydrogen-carbon     b. dust-air     c. sulfur-nitrogen     d. magnesium-iodine

2. Multiwire branch circuits are not permitted in Zone 20 locations.

   a. True     b. False

3. "Constructed so that the entrance of dust is limited under specified test conditions" is the definition of a dusttight enclosure.

   a. True     b. False

4. Electrical equipment suitable for ambient temperatures exceeding 30°C (86°F) shall be marked with both the maximum ambient temperature and the operating temperature at that ambient temperature.

   a. True     b. False

5. An area where combustible dust or ignitable fibers and flyings are likely to exist occasionally under normal operation in quantities sufficient to be hazardous is classified as a _____ Hazardous Location.

   a. Zone 22     b. Zone 20     c. Zone 21     d. Zone 23

# ● ARTICLE 510–HAZARDOUS (CLASSIFIED) LOCATIONS–SPECIFIC

No questions

# ● ARTICLE 511–COMMERCIAL GARAGES, REPAIR, AND STORAGE

1.  Plugs connections to an automobile must be arranged so that the lowest point of a sag is at least _____ above the floor of an automotive shop.

    a. 4'                 b. 12"                 c. 4"                 d. 6"

2.  Any unventilated pit or depression below floor level is considered to be Class I, Division 2 and must extend up to said floor level.

    a. True                 b. False

3.  Battery chargers and batteries being charged are permitted to be located in classified areas.

    a. True                 b. False

# ● ARTICLE 513–AIRCRAFT HANGARS

1. Lampholders of metal-shell, fiber-lined types shall not be used for fixed incandescent lighting.
   a. True                    b. False

2. An approved means must be provided for maintaining continuity of the grounding conductor between fixed wiring systems and non-current-carrying metal portions of pendant fixtures, portable lamps, and utilization equipment.
   a. True                    b. False

3. Equipment with electric components suitable to be moved by a single person without mechanical aids is defined as _____
   a. movable equipment
   b. portable equipment
   c. mobile equipment
   d. mechanical equipment

4. While undergoing maintenance and stored in a hangar, an aircraft's electrical system must be _____.
   a. locked out          b. tagged              c. inactive            d. shut off

5. The area within _____ from aircraft power plants or aircraft fuel tanks shall be classified as a Class I, Division 2 or Zone 2 location that shall extend upward from the floor to a level _____ about the upper surface of wings and of engine enclosures.
   a. 7' vertically, 5'    b. 10' below, 7'        c. 5' horizontally, 5'    d. 5' above, 10'

6. Mobile energizers shall carry at least one permanently affixed warning sign displaying which of the following words?
   I. WARNING KEEP 5 FT CLEAR OF AIRCRAFT ENGINES AND FUEL TANK AREAS
   II. WARNING KEEP 5 METERS CLEAR OF AIRCRAFT ENGINES AND FUEL TANK AREAS
   a. I only              b. II only             c. Both I & II          d. neither I nor II

# ● ARTICLE 514–MOTOR FUEL DISPENSING FACILITIES

1. Where a restroom in a service station can only be entered through a Division 1 location, the restroom shall be classified as _____.

   a. Class I, Division 1

   b. Class I, Group D, Division 1

   c. Class I, Group D, Division 2

   d. Class I, Division 2

2. _____ shall shut off all power to all dispensing equipment at unattended facilities.

   a. Disconnect switches

   b. Circuit breakers

   c. Emergency controls

   d. Remote controls

3. An area is not required to be classified when the authority having jurisdiction can satisfactorily determine that flammable liquids, such as gasoline, having a flash point below _____, will not be handled.

   a. 100°C          b. 38°F                c. 100°F                d. 38°C

4. The area classification for below-grade work area ventilated pits with a 1.5 cfm/ft² exhaust rate in a lubrication shop without dispensing capabilities is _____.

   a. Class I, Division 2          b. unclassified          c. Class I, Group D, Division 1

   d. Class I, Division 1

5. The area within _____ of the open end of an upward discharging vent is classified as a _____ location.

   a. 5'; Class I, Division 2          b. 3'; Class I, Division 1          c. 7'; Class I, Division 1

   d. 10'; Class I, Division 2

6. An approved seal must be provided in each conduit run entering or leaving a dispenser or any cavity or enclosure in direct communication therewith.

   a. True                b. False

# ⊙ ARTICLE 515–BULK STORAGE PLANTS

1.  From the area inside a dike to the top level of a tank is classified as a _____ location.
    a. Class II          b. Zone 0          c. Division 1          d. Zone 2

2.  Where storage lockers are used for the storage of Class I liquids, the enclosing of the lockers must be classified as a Class I, Division 2 Zone 2 location in its entirety.
    a. True          b. False

3.  Class I locations are permitted to be extended beyond a roof, floor, wall, or other solid partition that has no communicating openings.
    a. True          b. False

4.  Fixed wiring above Class I locations shall be in metal or nonmetallic Schedule 80 PVC conduit or be Type MI, TC, or MC cable.
    a. True          b. False

5.  If any part of a sump is within a Division 2 Zone 2 classified location, the entire area must be classified as a _____.
    a. Class I, Division 1 Zone 2          b. Class I, Division 1 Zone 1
    c. Class I, Division 1 Zone 1          d. Class I, Division 2 Zone 1

6.  Any area within 4' of the open end of a downward discharging vent must be classified Class I, Division 2.
    a. True          b. False

7.  An equipment grounding conductor must be included when rigid nonmetallic conduit or cable with a nonmetallic sheath is used to ensure electrical continuity of raceway systems for the grounding of non-current-carrying metal parts.
    a. True          b. False

# ● ARTICLE 516–SPRAY APPLICATION, DIPPING, AND COATING PROCESSES

1. In manual operations, the area limits shall be the maximum area of spray when aimed at 180° to the application surface. This brief description best describes a _____.

   a. spray room          b. spray area          c. spray booth          d. spray enclosure

2. Any area in the direct path of spray operations is classified as what type location?

   a. Class I, Division 1

   b. Class II, Division 1

   c. either Class I or Class II, Division 1

   d. None of the preceding

3. A 6' space above the floor is required for dip tanks and drain boards and must extend 15' horizontally in all directions from Class I, Division 1 locations.

   a. True          b. False

4. High-voltage grids, electrodes, electrostatic atomizing heads, and their connections shall be permitted within Class I locations.

   a. True          b. False

5. Electrostatic fluidized beds and associated equipment shall be of _____ types.

   a. approved          b. identified          c. listed          d. classified

6. Except those objects required by the process to be at high voltage, all electrically conductive objects in a spray area shall be adequately _____.

   a. protected against short circuits          b. grounded          c. covered          d. isolated

# ● ARTICLE 517–HEALTH CARE FACILITIES

1.  Receptacles and attachment plugs installed and used in other-than-hazardous locations shall be listed for hospital use; this applies to 3-phase, 4-wire multiwire circuits.

    a. True                 b. False

2.  In a general care area, each patient bed location must be supplied with at least two branch circuits: one for emergency loads and the other for critical loads.

    a. True                 b. False

3.  A rating based on an operating interval that does not exceed 5 seconds is defined as which of the following?

    a. X-ray installations, micro-time rating          b. X-ray installations, short-time rating
    c. X-ray installations, momentary rating          d. X-ray installations, flash-time rating

4.  Deleting the time-lag intervals feature for delayed automatic connections to the equipment system is permitted for essential electrical systems under _____.

    a. 5kW           b. 150kVA           c. 480V           d. 200A

5.  Routine housekeeping procedures and incidental spillage of liquids do not define a _____.

    a. hazardous location      b. wet location      c. classified area      d. laboratory

6.  An acceptable alternate means of providing isolation for patient/nurse call systems is by the use of nonelectrified signaling, communication, or control devices held by the patient or within reach of the patient.

    a. True                 b. False

7.  A hospital is considered a building or part thereof used for the 24-hour medical, psychiatric, obstetrical, or surgical care of _____ or more inpatients.

    a. 10           b. 15           c. 35           d. 4

8.  A minimum of _____ receptacles shall be provided for each patient bed in a general care area.

    a. 4           b. 2           c. 6           d. 8

9.  Where flammable anesthetics or volatile flammable disinfecting agents are stored, all rooms or location must be classified from floor to ceiling as _____.

    a. Class II, Division 1      b. Class I, Division 2      c. Class I, Division 1
    d. Class II, Division 2

10. The hazard current of a given isolated system with all devices, including the line isolation monitor connected, is defined as a _____.

    a. monitor hazard current      b. total hazard current      c. fault hazard current
    d. hazard current

11. Where operating over _____ in an area used for patient care, the grounding terminals of all receptacles and all non-current-carrying conductive surfaces of fixed electric equipment likely to become energized that are subject to personal contact shall be connected to an insulated copper _____ conductor.

    a. 125V, bare      b. 50V, stranded      c. 120V, green insulated      d. 115V, separate

    e. none of the preceding

12. _____ is the minimum size disconnecting means required for an X-ray machine if the machine has a momentary rating of 84A and a long-time rating of 35A.

    a. 30A              b. 60A              c. 100A              d. 150A

13. A line isolation monitor must alarm for a fault hazard of less than 3.7mA or for a total hazard current of less than 5mA.

    a. True              b. False

14. Single-phase fractional horsepower motors are permitted to be connected to the critical branch.

    a. True              b. False

15. Dining rooms and lounges located in a health care facility are classified as patient care areas.

    a. True              b. False

16. When the outside design temperature is higher than 20°F, heating of general patient rooms during disruption of normal power is not required.

    a. True              b. False

17. The ampacity of supply branch-circuit conductors and the current rating of overcurrent protective devices for diagnostic equipment shall not be less than _____ of the momentary rating or _____ of the long-time rating, whichever is greater.
    a. 100 percent, 125 percent
    b. 80 percent, 175 percent
    c. 125 percent, 80 percent
    d. 50 percent, 100 percent

18. _____ and limited care facilities that are located on the same site with a hospital shall be permitted to have their essential electrical systems supplied by that of the hospital.
    a. Inpatient care areas    b. Outpatient    c. Nursing homes    d. Day surgery

19. The equipment grounding terminal buses of the normal and essential branch-circuit panelboards serving the same individual patient care vicinity shall be connected together with an insulated continuous copper conductor not smaller than _____.
    a. No. 12 AWG    b. No. 8 AWG    c. No. 10 AWG    d. No. 6 AWG

20. Any area of a health care facility that is designated to be used for the administration of any flammable inhalation anesthetic agents in the normal course of examination or treatment is recognized as a flammable anesthetizing location.
    a. True    b. False

21. A ground-fault circuit interrupter is required in critical care areas where the toilet and basin are installed within a patient's room.
    a. True    b. False

22. Isolated circuits of an induction room shall be permitted to be supplied from the isolation _____ of any one of the operating rooms served by an induction room when the induction room serves more than one room.
    a. feeder    b. circuit    c. transformer    d. power

23. Cover plates for electrical receptacles or the electrical receptacles themselves supplied from the emergency system shall have a distinctive color or marking as to be _____ identified.
    a. clearly    b. visibly    c. readily    d. immediately

24. A rating based on an operating interval of 5 minutes or longer is best defined as which of the following?
    a. X-ray installations, short-time rating
    b. X-ray installations, long-time rating
    c. X-ray installations, momentary rating
    d. X-ray installations, intermediate rating

25. Individual branch circuits shall not be required for portable, mobile, and transportable medical X-ray equipment requiring a capacity of not over _____.

    a. 300W                 b. 50V                          c. 60A                          d. 300mA

26. Demand calculations for sizing generator set(s) are based on which of the following conditions?

    a. connected load

    b. feeder calculation procedures per Article 220

    c. prudent demand factors and historical data

    d. all of the preceding

# ⊙ ARTICLE 518–ASSEMBLY OCCUPANCIES

1.  Where fire-rated construction is not required, nonmetallic-sheathed cable, Type AC cable, electrical nonmetallic tubing, and rigid nonmetallic conduit is permitted to be installed in places of assembly.
    a. True           b. False

2.  The neutral conductor of feeders supplying solid-state, 3-phase, 4-wire dimmer systems shall not be considered a current-carrying conductor.
    a. True           b. False

3.  Places of assembly are designed to occupy _____ or more persons.
    a. 50        b. 100        c. 75        d. 200

# ⊙ ARTICLE 520–THEATERS, AUDIENCE AREAS OF MOTION PICTURE AND TELEVISION STUDIOS, PERFORMANCE AREAS, AND SIMILAR LOCATIONS

1.  Footlights, border lights, and proscenium sidelights shall be arranged so that no branch circuit supplying such equipment carries a load exceeding _____.
    a. 15A        b. 250W        c. 20A        d. 5kVA

2.  A circuit supplying an autotransformer-type dimmer shall not exceed _____ between conductors.
    a. 120V        b. 150V        c. 208V        d. 277V

3.  A hard usage cord must be restricted to _____ in overall length.
    a. 3'        b. 3.3'        c. 3.5'        d. 3.75'

4.  Cable or conductors that are physically tied, wrapped, taped, or otherwise periodically bound together are considered _____.
    a. grouped        b. secured        c. bounded        d. bundled

5.  Where the total length from supply to switchboard exceeds 100 feet, one additional interconnection shall be permitted for each additional _____ of supply conductor.
    a. 25 feet        b. 50 feet        c. 75 feet        d. 100 feet

6.  Curtain machines are required to be _____.
    a. approved        b. labeled        c. identified        d. listed

7.  An adapter cable containing one male plug and two female cord connectors used to connect two loads to one branch circuit is called a U-fer.
    a. True           b. False

# ● ARTICLE 522–CONTROL SYSTEMS FOR PERMANENT AMUSEMENT ATTRACTIONS

1.  Non-power limited control circuits shall not exceed _____.
    a. 300V                 b. 300A                 c. 300W                 d. 300Ω

2.  Where power or control circuits are in a metal-enclosed cable, control and power circuits are permitted to be installed as grounded conductors in a manhole.
    a. True                 b. False

3.  To provide an entertainment experience, entertainment devices may include animated props coordinated with audio and lighting.
    a. True                 b. False

4.  Single conductors _____ AWG or larger shall be permitted for jumpers and special wiring applications.
    a. 18          b. 24          c. 20          d. 16          e. answer not available

5.  The conditions of maintenance and supervision must ensure only that only personnel in the amusement industry can render service to permanent amusement attractions.
    a. True                 b. False

6.  Where the ampacity of a conductor with 90°C or greater insulation is adjusted, the conductor's ampacity must be based on 75°C.
    a. True                 b. False

# ● ARTICLE 525–CARNIVALS, CIRCUSES, FAIRS, AND SIMILAR EVENTS

1.  All metal frames and metal parts of rides connected to the same source must be
    _____.

    a. grounded          b. bonded          c. separately protected          d. isolated

2.  Where installed outside, termination boxes shall be of _____ construction.

    a. rainproof          b. watertight          c. raintight          d. weatherproof

3.  Single-conductor cable shall be permitted only in sizes _____ or larger.

    a. 8 AWG          b. 1 AWG          c. 2 AWG          d. 250 kcmil

4.  Amusement rides or attractions shall not be located under or within 15 feet horizontally of conductors operating in excess of 600 volts.

    a. True          b. False

# • ARTICLE 530–MOTION PICTURE AND TELEVISION STUDIOS AND SIMILAR LOCATIONS

1. Splices and taps are permitted where such are made with listed devices and the circuit is protected at not more than _____.

   a. 15A        b. 250W        c. 20A        d. 5kVA

2. Switchboards under _____ between conductors, where located in substations or switchboard rooms accessible to qualified persons only, shall not be required to be dead-front.

   a. 50VDC        b. 120VAC        c. 250VDC        d. 480VAC

3. A wall-mounted safety switch that is externally operated that may or may not contain overcurrent protection and that is designed for the connection of portable cables and cords is defined as a _____.

   a. disconnect switch        b. temporary switch        c. bull switch        d. circuit breaker

4. The overcurrent device setting for each feeder in buildings used primarily for motion picture production shall not exceed _____ of the feeder's ampacity based on the tables provided in Article 310.

   a. 80 percent        b. 125 percent        c. 250 percent        d. 400 percent

5. A DC device consisting of at least one 2-pole, 2-wire, nonpolarized, nongrounding-type receptacle intended to be used on DC circuits only is called a plugging box.

   a. True        b. False

6. A branch circuit of any size supplying two or more receptacles shall be permitted to supply stage set lighting loads.

   a. True        b. False

# ⦿ ARTICLE 540–MOTION PICTURE PROJECTION ROOMS

1. All project equipment must be _____.

   a. approved          b. listed          c. labeled          d. identified

2. Conductors supplying outlets for arc and xenon projectors of the professional type shall not be smaller than _____ and shall have an ampacity not less than the projector current rating.

   a. No. 14 AWG        b. No. 12 AWG       c. No. 10 AWG       d. No. 8 AWG

3. To control auditorium lights or switches for the control of motors operating curtains and masking of the motion picture screen, remote-control switches are permitted to be installed in projection rooms.

   a. True              b. False

4. A professional projector uses carbon arc, xenon, or other light source equipment that develops hazardous gases, dust, or radiation.

   a. True              b. False

# ⦿ ARTICLE 545–MANUFACTURED BUILDINGS

1. If the point of attachment is known, service-entrance conductors can be installed prior to erection at the building site.

   a. True              b. False

2. Provisions shall be made to route a grounding electrode conductor from the _____ supply to the point of attachment to the grounding electrode.

   a. service

   b. feeder

   c. branch-circuit

   d. any of the preceding choices

# ● ARTICLE 547–AGRICULTURAL BUILDINGS

1. _____ floors that are supported by structures that are a part of an equipotential plane shall not require bonding.

   a. Insulated       b. Sloped       c. Slatted       d. Concrete

2. For the purpose of equipotential planes, the term "livestock" does include domestic fowl.

   a. True       b. False

3. All 125V, single-phase, 15- and 20-ampere general-purpose receptacles installed _____ are required to be ground-fault protected.

   a. outdoors

   b. in areas having an equipotential plane

   c. in damp and wet locations

   d. all of the preceding choices

4. The center yard pole, meterpole, and common distribution points are also known as distribution points.

   a. True       b. False

# ⊙ ARTICLE 550–MOBILE HOMES, MANUFACTURED HOMES, AND MOBILE HOME PARKS

1. When an air conditioner is not installed in a mobile home and a 40A power-supply cord is provided, an allowance of 15A per leg should be considered for air conditioning.

   a. True           b. False

2. In mobile homes, GFCI protection is required for all receptacle outlets serving countertops in kitchens, and receptacle outlets located within 4 feet of a wet bar sink.

   a. True           b. False

3. Service equipment shall be located in sight of and not more than _____ from the exterior wall of the mobile being served.

   a. 10 ft.       b. 50 ft.       c. 30 ft.       d. 100 ft.

4. From the face of the attachment plug cap to the point where the power-supply cord enters a mobile home, the length of the cord must not be less than _____.

   a. 21'       b. 36'       c. 20'       d. 18'

5. Bonding conductors shall be solid or stranded, insulated, or bare, and shall be _____ minimum or equivalent.

   a. No. 8 aluminum     b. No. 6 copper     c. No. 8 copper     d. No. 6 aluminum

6. The square footage of a manufactured homes is based on the structure's interior dimensions measured at the largest horizontal projection when erected on site.

   a. True           b. False

7. Neither the frame of a mobile home nor the frame of any appliance can be connected to the _____ conductor in a mobile home.

   a. grounding electrode     b. equipment grounding     c. grounded     d. grounding

8. The unit load for determining the number of branch circuits for a mobile home is _____.

   a. 3.5 VAft$^2$       b. 3 VA/ft$^2$       c. 2.5 VA/ft$^2$       d. 4 VA/ft$^2$

9. When the calculated load of a mobile home is 14,500 volt-amperes, can the actual load calculation be used, or is it required to be increased to 16,000 volt-amperes?

   a. use actual calculation     b. increase to 16,000 volt-amperes     c. not applicable

10. Which of the following appliances is not considered portable?

    a. refrigerator          b. built-in microwave          c. washer          d. range

11. Mobile home and manufactured home lot feeder circuit conductors shall have adequate capacity for the loads supplied and shall be rated at not less than 60 amperes at 240/120 volts.

    a. True          b. False

12. The wiring of all mobile homes shall be subjected to a _____, dielectric strength test between live parts and the mobile home ground.

    a. 6-minute, 265-volt

    b. 1-minute, 900-volt

    c. 2-minute, 490-volts

    d. 3-minute, 330-volts

13. Where factory equipped with gas central heating equipment and cooking appliances, a listed mobile home power-supply cord rated for _____ is permitted.

    a. 50A          b. 45A          c. 40A          d. 60A

14. Applying the demand table associated with NEC 550.18(B)(5), if a 12.7kW electric range is provided with a mobile home, the demand load for the range can be reduced to _____.

    a. 12kW          b. 10kW          c. 80 percent of the nameplate rating          d. 8.4kW

# ⦿ ARTICLE 551–RECREATIONAL VEHICLES AND RECREATIONAL VEHICLE PARKS

1.  Where switches are used for lighting circuits, the rating of the switch must not be less than _____, with an operating voltage of 120–125 volts.

    a. 150VA        b. 1440W        c. 2HP        d. 10A

2.  The first 20 amperes of a load at 80 percent must be used as part of the formula to determine the voltage converter rating.

    a. True        b. False

3.  Grounded conductors are allowed to be used as equipment grounding conductors for recreational vehicles or equipment within the recreational vehicle park.

    a. True        b. False

4.  Waste pipes are considered grounded if bonded to a chassis.

    a. True        b. False

5.  Site supply equipment shall be located not less than _____ or more than _____ above the ground.

    a. 2'6", 6'        b. 3', 6'6"        c. 2', 6'7"        d. answer not provided

6.  For access to a generator, compartment doors are allowed to be equipped with a _____.

    a. remote control        b. combination lock        c. locking system        d. alarm monitor

7.  When the electrical supply for a recreational vehicle site has more than one receptacle, the calculated load must be computed per number of receptacles.

    a. True        b. False

8.  A 50A, 240/120V power supply assembly shall be used where _____ or more circuits are employed.

    a. 5        b. 1        c. 6        d. 2

9.   When the site supply equipment contains a 250/125V receptacle, the equipment must be marked to reflect which of the following?

   a. equipment's ampere rating

   b. procedure for inserting or removing plug

   c. aluminum conductors permitted

   d. only a and c

10.  A vehicular portable unit mounted on wheels and constructed with collapsible partial side walls that fold for towing by another vehicle and unfold at the campsite to provide temporary living quarters for recreational, camping, or travel use is recognized as a _____.

   a. mobile home        b. camping trailer        c. recreational vehicle        d. travel trailer

11.  The external power-supply assembly for a recreation vehicle is not permitted to be less than the calculated load.

   a. True                b. False

12.  Where located in wet locations or outside of a building switches, circuit breakers, receptacles, etc., shall be _____ equipment.

   a. weatherproof       b. raintight              c. waterproof                 d. rainproof

13.  The total demand load for a 33-site recreational vehicle park equipped with both 20 and 30 ampere supply facilities is approximately _____.

   a. 198kVA             b. 83kVA                  c. 119kVA                    d. 50kVA

14.  Where three receptacles are on a branch circuit, a 15A receptacle is permitted to be protected by a 20A overcurrent protection device.

   a. True                b. False

15.  The neutral conductors shall be permitted to be reduced in size below the minimum required size of the ungrounded conductors for _____, line-to-line, permanently connected loads only.

   a. 208V               b. 240V                   c. 250V                      d. 480V

16.  The point of first termination for supply conductors must be in a(n) _____.

   a. panelboard

   b. enclosed transfer switch

   c. junction box with a receptacle

   d. junction box with a blank cover

   e. all of the preceding

17. Ground-fault circuit interrupter protection is not required when a receptacle is located inside of an access panel that is installed on the exterior of an RV to supply power for an installed appliance.

    a. True                    b. False

18. A label measuring no less than _____ must be permanently affixed to the exterior skin of a recreational vehicle at or near the point of entrance of the vehicle's power-supply cord(s).

    a. 2" x 1 1/2"          b. 3" x 1.75"              c. 2" x 3"              d. 2" x 1 3/4"

19. Where electrical power is provided, all recreational vehicle sites must be equipped with at least two 20A, 125-volt receptacles.

    a. True                    b. False

# ⊙ ARTICLE 552–PARK TRAILERS

1. A low-voltage circuit consisting of No. 16 AWG stranded copper wire can be protected by a 15A overcurrent protective device.

   a. True            b. False

2. When an air conditioner is not installed in a mobile home and a 50A power-supply cord is provided, an allowance of 15A per leg should be considered for air conditioning.

   a. True            b. False

3. A park trailer is a unit that is built on a single chassis mounted on wheels and that has a gross trailer area not exceeding _____ in the set-up mode.

   a. 550 ft²      b. 625 ft²      c. 400 ft²      d. 375 ft²

4. An outlet used for pipe heating cable must be located with _____ of the cold water inlet.

   a. 2'      b. 18"      c. 4'      d. 36"

5. The rating of a single-cord and plug-connected appliance supplied by other than an individual branch circuit shall not exceed _____ of the circuit rating.

   a. 125 percent      b. 75 percent      c. 60 percent      d. 45 percent
   e. answer not provided

6. The distribution panelboard for park trailers shall be installed in a _____ location.

   a. fenced      b. readily accessible      c. dry      d. indoor

7. The operational test of low-voltage circuits shall be performed in the final stages of production after outer coverings and cabinetry have been _____.

   a. mounted      b. installed      c. secured      d. wrapped

8. Receptacle outlets are not permitted to be installed in a shower space.

   a. True            b. False

9. Applying the demand table associated with NEC 552.47(B)(5), if a 9.7kW electric range is provided with a mobile home, the demand load for the range can be reduced to _____.

   a. 9.2kW      b. 8.0kW      c. 7.76kW      d. 8.4kW

10. Bare conductors, green-colored conductors, or green conductors with white stripe(s) must be used for equipment-grounding conductors only.

    a. True                 b. False

11. How many 15A lighting circuits are required for a 25' x 16' park trailer?

    a. 3                b. 1                c. 2                d. no answer

# ● ARTICLE 553–FLOATING BUILDINGS

1.  Except at the service equipment of a floating building, the neutral conductor must be _____ from the equipment grounding conductor, equipment enclosures, and all other grounded parts.

    a. isolated          b. insulated          c. distinguished          d. separated

2.  A floating building's service equipment must be located adjacent to, but not in or on, the building or any floating structure.

    a. True              b. False

# ● ARTICLE 555–MARINAS AND BOATYARDS

1.  Where shore power accommodations provide two receptacles specifically for an individual boat slip and these receptacles have different voltage, only the receptacle with the _____ kilowatt demand shall be required to be calculated.

    a. average           b. least              c. highest              d. larger

2.  Receptacles that provide shore power for boats shall be rated not less than 30A and shall be duplex outlet type.

    a. True              b. False

3.  Yard and pier distribution systems shall not exceed _____ volts phase to phase.

    a. 208               b. 240                c. 480                  d. 600

4.  Cables and conductors must be routed to avoid wiring closer than 20 feet from the outer edge or any portion of the yard that can be used for _____.

    a. moving vessels

    b. stepping masts

    c. unstepping masts

    d. all of the preceding

5.  All electrical connections shall be located at least 18" above the deck of a floating pier.

    a. True              b. False

# ARTICLE 590—TEMPORARY INSTALLATIONS

1.  The equipment grounding conductor of all cord sets must be tested for _____ and shall be electrically continuous.

    a. shorts                  b. voltage drop              c. continuity              d. conductivity

2.  Receptacles are allowed to be connected to the same ungrounded conductor of multiwire circuits that supply temporary lighting.

    a. True                    b. False

3.  Temporary electrical power and lighting installations shall be permitted for a period not to exceed _____ for holiday decorative lighting and similar purposes.

    a. 1 year                  b. 6 months                  c. 8 weeks                 d. 90 days

# CHAPTER 6–SPECIAL EQUIPMENT

## ● ARTICLE 600–ELECTRIC SIGNS AND OUTLINE LIGHTING

1. In dry locations, the insulation on all conductors must extend not less than 2.5" beyond the metal conduit or tubing.

   a. True          b. False

2. Sign and outline lighting system equipment shall be at least _____ above areas accessible to vehicles unless protected from physical damage.

   a. 10'      b. 12'6"      c. 14'      d. 17'

3. Unless receptacles are provided, where electrodes penetrate an enclosure, only bushings listed for such purpose shall be used.

   a. True          b. False

4. For each sign and outline lighting system, a(n) _____ controlled operable switch or circuit breaker must be used to open all ungrounded conductors.

   a. separately      b. externally      c. remote      d. dedicated

5. Where installed in PVC, the length of the secondary circuit conductors from a high-voltage terminal to the first neon tube electrode shall not exceed _____.

   a. 15'      b. 32'      c. 50'      d. 72'

6. Equipment having an open circuit voltage exceeding _____ shall not be installed in or on dwelling occupancies.

   a. 480V      b. 550V      c. 750V      d. 1000V

7. A 20A branch circuit required for commercial buildings or occupancy shall supply no other loads.

   a. True          b. False

8. Transformers and electronic power supplies shall have a secondary-circuit current rating of not more than _____.

   a. 1A      b. 250A      c. 300 mA      d. .3 mA

9. Transformers with isolated ungrounded secondaries and with a maximum open circuit voltage of 7000 volts or less shall have secondary-circuit ground-fault protection.

   a. True          b. False

10. Where required for an electric sign, all bonding conductors must be copper and not smaller than _____ AWG.

   a. No. 14           b. No. 12           c. No. 10           d. No. 8

# ARTICLE 604–MANUFACTURED WIRING SYSTEMS

1. Connectors and receptacles shall be of the locking type, uniquely _____ and identified for the purpose, and shall be part of a listed assembly for the appropriate system.

   a. stamped          b. marked           c. polarized        d. labeled

2. Manufactured wiring systems are permitted to be used in outdoor locations where listed for such use.

   a. True             b. False

# ARTICLE 605–OFFICE FURNISHINGS
# (CONSISTING OF LIGHTING ACCESSORIES AND WIRED PARTITIONS)

1. Individual groups or partitions of interconnected individual partitions shall not contain more than _____ 15A, 125V receptacle outlets.

   a. 7                b. 10               c. 13               d. 5

2. Where cord and plug connections are provided for lighting equipment associated with wired partitions, the length of the cord must not exceed _____.

   a. 6'               b. 3'               c. 9'               d. 2'

3. Where permitted by the authority having jurisdiction, relocatable wired partitions shall be permitted to extend to the ceiling but shall not penetrate the ceiling.

   a. True             b. False

# ● ARTICLE 610–CRANES AND HOISTS

1. Taps without _____ overcurrent protection shall be permitted to brake coils.

   a. multiple          b. separate          c. dual          d. motor

2. Where a crane or hoist is operated by more than _____ motor(s), a common-return conductor of proper ampacity is permitted.

   a. three          b. five          c. one          d. six

3. A momentary switch or other device shall be provided to prevent the load block from passing the safe upper limit of travel of all hoisting mechanisms.

   a. True          b. False

4. When five cranes are supplied by a common conductor system, a demand factor of _____ must be applied.

   a. .75          b. .84          c. .78          d. .87

5. The continuous rating of the switch or circuit breaker used for disconnecting means shall not be less than _____ of the combined short-time ampere rating of the motors or less than _____ of the sum of the short-time ampere rating of the motors required for any single motion.

   a. 75 percent, 50 percent          b. 60 percent, 120 percent          c. 50 percent, 75 percent
   d. 80 percent, 35 percent

6. Where permitted, conductors external to motors and controls can be as small as _____ for electronic circuits.

   a. No. 18 AWG

   b. No. 20 AWG

   c. No. 16 AWG

   d. No. 14 AWG

   e. none of the preceding

7. Contact conductors can be used as feeders for any equipment other than crane(s) or hoist(s) that they are primarily designed to serve.

   a. True          b. False

8.  The nameplate of a monorail must only be marked to reflect the equipment's voltage and circuit amperes.

    a. True                          b. False

9.  One controller shall be permitted to be switched between motors, providing _____.

    I. only one motor is operated at one time

    II. the controller has a horsepower rating not lower than horsepower rating of the largest motor

    a. I                    b. II                    c. either I or II                    d. both I and II

10. Contact conductors are required to be enclosed in raceways.

    a. True                          b. False

11. Individual motor overload protection is not required for hoist and their trolleys that are not used as part of an overhead traveling crane, providing the largest motor does not exceed _____ and all motors are under manual control of the operator.

    a. 3 HP                    b. 5 HP                    c. 7.5HP                    d. 10HP

12. If two hoist motors were connected to the same branch circuit, the tap conductors to each motor are required to have an ampacity no less than _____ that of the branch circuit rating.

    a. 1/2

    b. 1/4

    c. 1/3

    d. 1/8

    e. not applicable

13. When a contact conductor is located 37' between end strain insulators, the conductor can be no smaller than a No. _____ AWG.

    a. 1                    b. 10                    c. 4                    d. .8                    e. 6

# ⦿ ARTICLE 620–ELEVATORS, DUMBWAITERS, ESCALATORS, MOVING WALKS, PLATFORM LIFTS, AND STAIRWAY CHAIR LIFTS

1. At least one 125-volt, single-phase, 15 or 20A _____ receptacle shall be provided in each machine room or control room and machinery space or control space.

   a. single      b. switched      c. duplex      d. ground-fault protected

2. For permanently installed sump pumps, a single receptacle is not required _____ protection.

   a. overcurrent

   b. ground-fault circuit interrupter

   c. overload

   d. arc-fault circuit interrupter

3. For heating and air-conditioning equipment located on the elevator car, branch circuits shall not have a circuit voltage in excess of 600 volts.

   a. True      b. False

4. Vertical runs of wireways shall be securely supported at intervals not exceeding _____.

   a. 4.5'      b. 8'      c. 12'      d. 15'

5. Providing that the ampacity is equivalent to at least a No. 14 AWG copper conductor, lighting circuits are allowed to be run in parallel with _____ or larger conductors.

   a. No. 18 AWG aluminum

   b. No. 20 AWG copper

   c. No. 16 AWG copper-clad aluminum

   d. none of the preceding

6. That length of cable, as measured from the point of suspension in the hoistway to the bottom of the loop, with the elevator car located at the bottom landing, is considered the unsupported length for a hoistway's suspension means.

   a. True      b. False

# ● ARTICLE 625–ELECTRIC VEHICLE CHARGING SYSTEM

1. The overall length of an electric vehicle supply equipment cable must not exceed _____ unless equipped with a listed cable management system.

   a. 6'           b. 16'                    c. 25'                    d. 32'

2. A 50A, 3-phase circuit that is used to charge electric 600Y/347V vehicles must provide a minimum ventilation of _____ cfm.

   a. 1066         b. 740                    c. 2135                   d. 854

3. Disconnecting means shall be provided and installed in a readily accessible location for electric vehicle supply equipment rate more than _____ amperes or more than _____ volts to ground.

   a. 150, 60      b. 30, 60        c. 100, 120       d. 60, 150       e. either a or d

4. Overcurrent protection for feeders and branch circuits supplying electric vehicle supply equipment is not required to be sized for continuous duty.

   a. True         b. False

# ● ARTICLE 626–ELECTRIFIED TRUCK PARKING SPACES (NEW)

1.  Unless equipped with a cable management system that is listed as suitable for the purpose, the overall length of the power-supply cord shall not exceed _____.
    a. 15'          b. 6'                    c. 18'                    d. 25'

2.  All electrified truck parking space supply equipment shall be accessible by an unobstructed entrance not less than _____ wide and not more than _____ high.
    a. 6'7", 2'          b. 2'6", 7'                    c. 2', 6.5'                    d. 2', 6'6"

3.  Conductors supplying truck space branch-circuit must have an ampacity not less than the load.
    a. True          b. False

4.  Electrified truck parking space single-phase branch circuits shall be derived from a _____.
    a. 240/120V, 3φ, 4-wire system          b. 480/277V, 3φ, 4-wire system
    c. 600/346V, 3φ, 4-wire system          d. 208/120V, 3φ, 4-wire system

5.  The power supply cable assembly shall be listed and rated for either 30A/480V/3φ or 60A/208V/3φ.
    a. True          b. False

6.  The disconnecting means shall be readily accessible and not located more than _____ from the receptacle it controls.
    a. 6"          b. 2.5'                    c. 18"                    d. 25'

7.  An electrified truck parking space also includes access roads and commercial parking areas.
    a. True          b. False

8.  Where an electrical supply exist, all truck parking space must be equipped with _____.

    I. two 20A/125V 1-gang receptacles          II. one 30A/208Y/120V receptacle
    III. one 30A 125/250V receptacle
    a. I                          b. II and III          c. I and II          d. II
    e. I and III                  f. III                 g. I, II and III     h. I and II or III
    i. I or II and III

9.  Where the climatic temperature zone is 6a, the demand factor for services and feeders is _____.
    a. 62 percent          b. 21 percent                    c. 39 percent                    d. 70 percent

# ⊙ ARTICLE 630–ELECTRIC WELDERS

1. The current drawn from the supply circuit during each welder operation at the particular heat tap and control setting used is called the _____.

   a. rated current      b. full-load current      c. actual primary current      d. cycle current

2. At intervals not greater than 20 feet, a permanent sign shall be attached to cable tray containing welding cable.

   a. True                b. False

3. The load value used for each welder considers both the magnitude and the duration of the load while the welder is in use.

   a. True                b. False

4. Insulation of conductors intended for use in the secondary circuit of electric welders must be _____.

   a. heat resistant      b. flame retardant      c. fireproof      d. hard-rubber coated

5. Connecting the secondary circuit of a welder to grounded objects can create parallel paths and can cause objectionable current over _____ conductors.

   a. grounded      b. equipment grounding      c. ungrounded      d. grounding electrode

# ⦿ ARTICLE 640–AUDIO SIGNAL PROCESSING, AMPLIFICATION, AND REPRODUCTION EQUIPMENT

1. The _____ wires of an audio transformer or autotransformer shall be allowed to connect directly to the amplifier or loudspeaker terminals.

   a. input          b. output          c. input and output          d. input or output

2. Cables and mats where accessible to the public must not be arranged to present a(n) _____ hazard.

   a. electrical          b. shock          c. tripping          d. public

3. When a wireway does not enclose power-supply wires, the equipment-grounding conductor shall not be required to be larger than _____ or its equivalent.

   a. No. 12 copper

   b. No. 10 copper

   c. No. 18 copper

   d. No. 14 copper

4. Equipment that combines the functions of a mixer and amplifier within a single enclosure is called a mixer.

   a. True          b. False

5. Audio amplifier output circuits wired using Class 1 wiring methods shall be considered equivalent to _____ circuits.

   a. Class 1 or 2          b. Class 2          c. Class 1          d. Class 1 and Class 3

6. A transformer with two or more electrically isolated windings and multiple taps intended for use with an amplifier loudspeaker signal output is called an audio autotransformer.

   a. True          b. False

# ● ARTICLE 645–INFORMATION TECHNOLOGY EQUIPMENT

1.  In an information technology equipment room, a disconnecting means is required to _____.

    a. disconnect the power to all electronic equipment only

    b. disconnect the power to all dedicated HVAC systems only

    c. neither a nor b

    d. both a and b

2.  Where the area under raised floors is accessible, interconnecting cables are permitted to be installed under raised floors.

    a. True                    b. False

3.  Where supplied by a branch circuit, each unit of an information technology system shall have a nameplate that includes _____.

    a. frequency

    b. maximum rated amps

    c. input power requirements for voltage

    d. all of the preceding

4.  Branch circuit conductors supplying one or more units of a data processing system shall have an ampacity not less than _____ of the total connected load.

    a. 80 percent          b. 125 percent          c. 100 percent          d. 150 percent

# ● ARTICLE 647–SENSITIVE ELECTRONIC EQUIPMENT

1. For cord-connected equipment, the combined voltage drop of feeder and branch-circuit conductors must exceed 5 percent.

   a. True            b. False

2. Where luminaires are installed to reduced electrical noise, they shall not have an exposed lamp screw-shell.

   a. True            b. False

3. The technical equipment grounding bus must be bonded to the enclosure of a panelboard.

   a. True            b. False

4. All junction box covers shall be clearly marked to indicate the distribution panel and the

   _____.

   a. short-circuit current

   b. system voltage

   c. kVA rating

   d. type feeder conductors

# ● ARTICLE 650–PIPE ORGANS

1. A main common-return conductor in the electromagnetic supply shall not be less than _____ AWG.

   a. No. 18      b. No. 12      c. No. 14      d. No. 16

2. An overcurrent device rated at not more than _____ must be used to protect 26 and 28 AWG conductors.

   a. 1A      b. 2A      c. 4A      d. 6A

3. The source of power for an electrically operated pipe organ must be a transformer-type rectifier, where the DC potential of which shall not exceed _____.

   a. 18VDC      b. 30VDC      c. 9VDC      d. 12VDC

# ● ARTICLE 660–X-RAY EQUIPMENT

1.  Battery operated X-ray equipment is required to have an approved grounding-type attachment plug cap.

    a. True                    b. False

2.  All new or reconditioned X-ray equipment moved to and reinstalled at a new location must be of a(n) _____ type.

    a. listed          b. approved          c. equivalent          d. recognized

3.  Stationary X-ray equipment properly supplied by a branch circuit rated at not over _____ shall be permitted to be supplied through a suitable attachment plug cap and hard-service cable or cord.

    a. 125 volts          b. 1440 volt-amperes          c. 30 amperes          d. 1000 watts

# ● ARTICLE 665–INDUCTION AND DIELECTRIC HEATING EQUIPMENT

1.  For capacitors rated over 600 volts, a means shall be provided to reduce the residual voltage of a capacitor to _____ or less within _____ after the capacitor is disconnected from the source of supply.

    a. 60V, 3 minutes          b. 15V, 5 minutes          c. 100V, 7.5 minutes          d. 50V, 5 minutes

2.  Switches operated by foot pressure shall be provided with a shield over the contact button to avoid accidental closing of a limit switch.

    a. True                    b. False

3.  The heating equipment applicator is the device used to _____ energy between the output circuit and the object or mass to be heated.

    a. store

    b. intercept

    c. separate

    d. impede

    e. none of the preceding

# ● ARTICLE 668–ELECTROLYTIC CELLS

1. Where isolating transformers are used to supply power for receptacles for portable electrical equipment, the transformer's primary voltage between conductors must not exceed _____ volts.

   a. 120               b. 208               c. 480               d. 600

2. General-purpose electrical equipment enclosures shall be permitted where a natural draft ventilation system prevents the accumulation of _____.

   a. vapor             b. condensation      c. gases             d. corrosion

3. The cell line working zone shall not be required to extend through or beyond walls, floors, roofs, partitions, barriers, or the like.

   a. True              b. False

4. The portion of a hoist that contacts an energized electrolytic cell or energized attachments shall be _____ from ground.

   a. isolated          b. separated         c. insulated         d. protected

5. Partial or total shunting of cell line circuit current around one or more cells is not permitted.

   a. True              b. False

# ● ARTICLE 669–ELECTROPLATING

1. The term "electroplating" must be used to identify the process of anodizing.

   a. True              b. False

2. Removable links are not permitted to be used as disconnecting means.

   a. True              b. False

# ● ARTICLE 670–INDUSTRIAL MACHINERY

1.  When furnished as a part of the machine, overcurrent protection shall consist of
    _____ .

    a. a single circuit breaker    b. a set of fuses    c. both a and b    d. either a or b

2.  The disconnecting means for industrial machinery is not required to be equipped with
    overcurrent protection.

    a. True                b. False

# ● ARTICLE 675–ELECTRICALLY DRIVEN OR CONTROLLED IRRIGATION MACHINES

1.  When several irrigation motors are on one branch circuit, the full-load current of any motor
    must not exceed 5A.

    a. True                b. False

2.  Feeder circuits supplying power to irrigation machines shall have an equipment grounding
    conductor sized according to _____ .

    a. Table 250.66    b. Table 250.122    c. Table 310.16    d. Table 430.52

3.  An assembly of slip rings for transferring electrical energy from a stationary to a rotating
    member is called a commutator.

    a. True                b. False

4.  When involving irrigation machines, which of the following pieces of equipment is
    required to be grounded?

    a. all electrical equipment on the irrigation machine

    b. metal enclosures and junction boxes

    c. associated control panels or control equipment

    d. all of the preceding

5.  The conductors of a cable used to interconnect enclosures on the structure of an irrigation
    machine shall be of a type suitable for an operating temperature of _____ .

    a. 75°F        b. 30°C        c. 167°F        d. 70°C

# ⦿ ARTICLE 680–SWIMMING POOLS, FOUNTAINS, AND SIMILAR INSTALLATIONS

1. Ground-fault protection is required for a field assembled spa or hot tub when a heater load exceeds 50A.

   a. True                    b. False

2. Wall-mounted light fixtures shall not be installed less than _____ below the normal water level of the pool.

   a. 6"            b. 12"            c. 4"            d. 18"

3. A duplex receptacle must not be installed within _____ of the inside walls of a pool.

   a. 7'            b. 6'            c. 14'            d. 17'

4. If located more than 5 feet from a pool, the electric motor and controller used to cover the pool is not required to be connected to a circuit protected by a GFCI.

   a. True                    b. False

5. Type B, liquidtight flexible non-metallic conduit is allowed to be longer than _____ when connecting transformers for pool lights.

   a. 3'            b. 4'            c. 5'            d. 6'

6. Without causing damage to building structure or finish, electrical equipment for hydromassage bathtubs must be accessible.

   a. True                    b. False

7. A hot tub is not designed to have its contents _____ after each use.

   a. cleaned        b. sanitized        c. drained        d. tested

8. Electrical metallic tubing is an approved wiring method for the installation of wet-niche lighting fixtures.

   a. True                    b. False

9. When a lighting outlet is located over an indoor spa or hot tub or within 5' from the inside walls of either, _____.

   a. GFCI protection is not required if mounting height is greater than 12'

   b. the mounting height of the outlet must not be less than 7.5', if GFCI protected

   c. either a or b

   d. both a and b

10. Which of the following is not considered a fountain?

    a. drinking fountain     b. display pools     c. reflection pools     d. ornamental pools

11. A 15A, 208V, single-phase receptacle that supplies pool pump motors must be GFCI protected.

    a. True            b. False

12. Submersible pumps must operate at _____ or less between conductors.

    a. 120V          b. 208V          c. 240V          d. 300V

13. The ampacity of branch circuit conductors and overcurrent protection for electric pool water heaters must not be less than _____ of the total nameplate rated load.

    a. 80 percent      b. 150 percent      c. 125 percent      d. 100 percent

14. As a minimum, the bonded parts of a pool are required to be connected to the equipotential bonding grid with what type conductor?

    a. No. 8 bare      b. No. 8 insulated      c. No. 8 covered      d. choices a–c

15. Electrical equipment located within 10' of the inside walls of a swimming pool is not required to be grounded.

    a. True            b. False

16. Where rated _____ or less at 125 volts, receptacles located within 10' of the inside walls of a spa or hot tub must be protected by a ground-fault circuit interrupter.

    a. 15A          b. 30A          c. 20A          d. 25A

17. Where located 5' from _____, electrical controls and devices that are not associated with such units are required to be bonded.

    a. fountains      b. spas      c. therapeutic tubs      d. sump pumps

18. Radiant heating cables for a pool deck _____.

    a. is required to be installed 5 feet horizontally from the inside walls of a pool

    b. shall be mounted at least 12 feet vertically above the pool deck

    c. either a or b

    d. both a and b

    e. is not permitted

19. Equipment that is not easily moved from one place to another in normal use is considered fixed.

    a. True                    b. False

20. Where wet-niche lighting fixtures are supplied by flexible cord, the cord's grounding conductor can be no smaller than _____.

    a. the supply conductors

    b. No. 16 AWG

    c. both a and b

    d. neither a nor b

21. If ground fault protected, portable electric signs can be placed inside fountains.

    a. True                    b. False

22. A cord and plug connection is permitted if protected by a ground-fault circuit interrupter and the cord is no longer than 10 feet.

    a. True                    b. False

23. The smallest sized copper equipment grounding conductor allowed for motors that are associated with permanently installed swimming pools is _____.

    a. No. 10          b. No. 8 copper          c. No. 6 copper          d. correct choice not given

## ⦿ ARTICLE 682–NATURAL AND ARTIFICIALLY MADE BODIES OF WATER

1. A _____ is the farthest extent of standing water under the applicable conditions that determine the electrical datum plane for the specified body of water.

   a. river bank      b. stream      c. shoreline      d. retention basin

2. Circuits rated not more than 60A at 120-250V, single phase, shall have GFCI protection.

   a. True      b. False

3. Equipment grounding conductors shall be insulated aluminum conductors not smaller than 12 AWG.

   a. True      b. False

4. Service equipment shall disconnect when the water level reaches the height of the established datum plane.

   a. True      b. False

## ⦿ ARTICLE 685–INTEGRATED ELECTRICAL SYSTEMS

1. Two-wire DC circuits are required to be grounded.

   a. True      b. False

2. An integrated electrical system is a(n) _____ segment of an industrial wiring system.

   a. integrated      b. intermixed      c. unitized      d. partial

# ◉ ARTICLE 690–SOLAR PHOTOVOLTAIC SYSTEMS

1. A plaque or directory is required in a building to indicate the location of a photovoltaic power system's disconnecting means.

   a. True          b. False

2. When determining the rating of the overcurrent protection for a photovoltaic system, the utilization current is considered continuous.

   a. True          b. False

3. A set of modules interconnected as a system must be considered a single-source circuit when _____.

   a. it either is with or without blocking diodes          b. it is rated for 50 volts or less

   c. it has a single overcurrent device          d. all of the preceding

4. The basic photovoltaic device that generates electricity when exposed to light is called a _____.

   a. fiber optic          b. solar cell          c. photo-electric emitter          d. photo cell

5. In sizes _____ and larger, flexible cables shall be permitted within battery enclosures from battery terminals to a nearby junction box, where they shall be connected to an approved wiring method.

   a. No. 8 AWG          b. No. 4 AWG          c. No. 1 AWG          d. No. 2/0 AWG

6. The storage batteries for dwellings must have the cells connected so as to operate at less than _____, nominal.

   a. 35V          b. 60V          c. 48V          d. 50V

7. Equipment that controls DC voltage or AC current, or both, used to charge a battery is called a charge controller.

   a. True          b. False

8. Type(s) _____ and single-conductor cable listed and labeled as photovoltaic (PV) wire shall be permitted in exposed outdoor locations in photovoltaic source circuits for photovoltaic module interconnections within the photovoltaic array.

   a. USE          b. SE          c. UF          d. USE-2          e. all choices

9. A power transformer with a current rating on the side connected toward the _____, not less than the short-circuit output current rating of the inverter, shall be permitted without overcurrent protection from that source.

   a. inverter output          b. photovoltaic power source          c. module          d. panel

# ARTICLE 692–FUEL CELL SYSTEMS

1. The term "compatible" means matching the primary source wave shape.

   a. True                b. False

2. A fuel cell system that supplies power independently of an electrical production and distribution network is called a(n) _____.

   a. independent fuel cell system      b. fuel cell system      c. stand-alone system

   d. primary power source

3. The manual fuel shut-off valve shall be at the location of the _____ disconnecting means of the building or circuits applied.

   a. service          b. primary          c. main          d. power supply

4. The highest fuel cell inverter output voltage between any ungrounded conductors present at accessible output terminals is called the peak system voltage.

   a. True                b. False

# ● ARTICLE 695–FIRE PUMPS

1. As an approved wiring method, nonmetallic rigid conduit can be installed between controller(s) and pump motors.

   a. True                b. False

2. Generator control conductors must have a minimum _____ fire resistance rating.

   a. 30-minute       b. 1-hour             c. 1.5-hour           d. 3-hour

3. A fire pump is permitted to be supplied by _____.

   a. a separate service

   b. a tap located ahead of and not within the same cabinet, enclosure, or vertical switchboard section as the service disconnecting means

   c. either a or b

   d. both a and b

   e. none of the preceding

4. A fire pump's controller and power transfer switch can serve any load that is associated with a fire pump.

   a. True                b. False

5. A single disconnecting means with overcurrent devices is permitted to be installed between a remote power source and a listed fire pump power transfer switch.

   a. True                b. False

6. When operating at 115 percent of a motor's full-load current rating, the voltage at the motor's terminals must not fall below _____ of the motor's voltage rating.

   a. 3 percent       b. 5 percent           c. 7 percent           d. 10 percent

# CHAPTER 7–SPECIAL CONDITIONS

⦿ **ARTICLE 700–EMERGENCY SYSTEMS**

1.  When a combustion engine is used as the prime mover of a generator set used to supply emergency power, a _____ on-site fuel supply is required.

    a. 3-hour          b. 1/2-hour          c. 5-hour          d. 2-hour

2.  When a building used for assembly exceeds _____ persons, equipment for feeder circuits must be located in a space with a 1-hour fire resistance rating.

    a. 100          b. 300          c. 700          d. 1000

3.  Where practical, audio and visual signal devices shall be provided for the purpose of _____.

    a. indicating battery is carrying load

    b. indicating derangement of emergency source

    c. indicating battery charger is not functioning

    d. indicating ground fault in solidly grounded wye systems

    e. all of the preceding choices

4.  Transfer equipment, along with automatic transfer switches, is required to be _____.

    a. identified for emergency use

    b. approved by the authority having jurisdiction

    c. automatic

    d. all of the preceding choices

5.  A control switch for emergency lighting in theaters, motion-picture theaters, or places of assembly under no conditions shall be placed _____.

    a. on a platform

    b. on a stage

    c. in a motion-picture projection booth

    d. all of the preceding

6.  Three and four-way switches are not allowed to be used for emergency lighting circuits.

    a. True          b. False

⦿

7.  The authority having jurisdiction must be provided with an acceptable schedule to ensure emergency systems are periodically tested and maintained in proper operating conditions.

    a. True                 b. False

8.  For a generator set driven by a prime mover, a time-delay feature permitting a _____ setting must be provided to avoid retransfer in the event of short-time reestablishment of the normal source.

    a. 1.5-minute          b. 6-minute              c. 12-minute             d. 15-minute

9.  Storage batteries used as a source of power for emergency systems must have the capacity to supply and maintain the total load for a period of _____ without the voltage applied to the load falling below _____.

    a. 1/2 hour, 95 percent

    b. 1 hour, 90 percent

    c. 1 3/4 hours, 75 percent

    d. 2 hours, 60 percent

    e. none of the preceding

10. Buildings occupying emergency systems where daylight exterior lighting is not required are allowed to use an automatic light-actuated device to control lighting when needed.

    a. True                 b. False

# ● ARTICLE 701–LEGALLY REQUIRED STANDBY SYSTEMS

1.  Branch-circuit overcurrent devices in legally required standby circuits shall be accessible to authorized persons only.

    a. True                 b. False

2.  A sign must be placed _____ indicating type and location of on-site legally required standby power sources.

    a. on the generator set

    b. at the service entrance

    c. within the monitored area

    d. by maintenance personnel

3.  Legally required systems are typically not installed to serve _____ loads.

    a. sewage disposal      b. operating rooms          c. HVAC systems          d. lighting

4.  Where water is required to be added to lead acid batteries, transparent or translucent jars can be used.

    a. True                 b. False

# ● ARTICLE 702–OPTIONAL STANDBY SYSTEM

1.  When the intention is to supply private businesses or property where life safety does not depend on the performance of the system, the system is described as a(n) _____.

    a. emergency system

    b. optional system

    c. legally required standby system

    d. UPS system

2.  Where manual transfer equipment is used, only those personnel servicing an optional standby system are allowed to select the load connected to the system.

    a. True                 b. False

# ⊙ ARTICLE 705–INTERCONNECTED ELECTRIC POWER PRODUCTION SOURCES

1.  A generator's output or other electric power production source operating in parallel with an electric supply system shall be compatible with _____ of the system to which it is connected.

    a. the frequency

    b. the voltage

    c. the wave shape

    d. all of the preceding

2.  Synchronous generators in a _____ system shall be provided with the necessary equipment to establish and maintain a synchronous condition.

    a. serial                 b. bypass                 c. parallel                 d. time-delayed

3.  Installation with large numbers of power production sources is not allowed to be designated by groups.

    a. True                 b. False

4.  When equipment is intended to be operated and maintained as an integral part of a power production source, a disconnecting means is not required when the voltages exceeds

    _____.

    a. 300V                 b. 480V                 c. 750V                 d. 1000V

5.  The _____ of electric power production systems shall be interconnected at the premise's service disconnecting means.

    a. service drop

    b. watt-hour meter

    c. service lateral

    d. overcurrent protection

    e. none of the preceding

6.  The output of an interactive system shall be connected to the load side of the ground-fault current sources when ground-fault protection is used.

    a. True                 b. False

# ARTICLE 708–CRITICAL OPERATIONS POWER SYSTEMS (COPS) (NEW)

1. Feeder COPS equipment shall be located _____ the 100-year floodplain.

   a. adjacent to          b. beneath          c. above          d. within

2. A 5-cycle minimum separation between the service and feeder ground-fault tripping bands shall be provided.

   a. True          b. False

3. When maintenance and testing of COPS is conducted, records must be kept.

   a. True          b. False

4. An area within a facility or site designated as requiring critical operations power is a _____.

   a. critical operations area          b. supervisory control area
   c. risk assessments area             d. designated critical operations area

5. An alternate power source must be capable of maintaining a steady-state voltage within _____ of the nominal utilization voltage where critical operations power systems are operational.

   a. 10 percent          b. 5 percent          c. +/- 10 percent          d. +/- 5 percent

6. A time-delay feature permitting a minimum _____ setting shall be provided to avoid retransfer in case of short-time reestablishment of the normal source.

   a. 2 minutes     b. 5.5 seconds     c. 10 minutes     d. 1/2 hour     e. answer not provided

7. In a critical operations power system, risk assessment shall be performed only to identify hazards.

   a. True          b. False

# ⦿ ARTICLE 720–CIRCUITS AND EQUIPMENT OPERATING AT LESS THAN 50 VOLTS

1. Where portable appliances are likely to be used, receptacles having a rating not less than 15A shall be provided in kitchens, laundries, and other locations.

   a. True　　　　　　　　b. False

2. Standard lampholders that have a rating of not less than _____ shall be used.

   a. 7.5A　　　　　　b. 250V　　　　　　c. 660W　　　　　　d. 150VA

3. For appliance branch circuits supplying more than one appliance or appliance receptacle, branch circuit conductors must not be smaller than No. 12 AWG.

   a. True　　　　　　　　b. False

# ARTICLE 725–CLASS 1, CLASS 2, AND CLASS 3 REMOTE-CONTROL, SIGNALING, AND POWER-LIMITED CIRCUITS

1. When used in hoistways, Class 2 and 3 circuit conductors shall be installed in _____.

   a. electrical metallic tubing

   b. intermediate metal conduit

   c. flexible nonmetallic conduit

   d. all of the preceding

2. Class 1 circuits in sizes 18 and 16 AWG must use conductors with insulation types _____.

   I. PAF, FFH-2 KF-2      II. RFH-2, PGF, RFHH-3      III. ZW, THW, XHHW

   a. I only               b. II only                 c. III only            d. only I and II

3. For type PLTC nonmetallic sheathed, power-limited tray cable, the cable core must be _____.

   a. two or more parallel conductors

   b. a combination thereof

   c. one or more group assemblies of twisted or parallel conductors

   d. none of the preceding

   e. all of the preceding

4. The limits of the output current ($I_{max}$) applies after _____ when a current-limiting impedance is used in combination with a stored energy.

   a. 1 minute             b. 5 seconds               c. 3 1/2 minutes       d. 1 hour

5. A dry cell battery shall be an inherently limited Class 2 power source, provided the voltage is _____ and the capacity is equal to or less than that available from series connected _____ carbon zinc cells.

   a. 6 volts or less, No. 30

   b. 12 volts or more, No. 9

   c. 30 volts or more, No. 6

   d. 9 volts or more, No. 12

6. Which of the following type cable are permitted for Class 2 and 3 circuits?

   a. CL2P                 b. CL3P                    c. either a or b       d. both a and b

7.  Voltage markings on cables may be misinterpreted to suggest that the cables may be suitable for Class 1 electric light and power applications.

    a. True                    b. False

8.  Class 1 and power circuits are permitted to occupy the same cable, enclosure, or raceway only when the equipment powered is functionally associated.

    a. True                    b. False

9.  Where listed for use, Type _____ cables are permitted to be directly buried.

    a. CL2               b. PTLC               c. CL3X               d. CL2R

10. _____ circuits consider safety from a fire initiation standpoint due to their power limitations.

    a. Class 1          b. Class 2          c. Class 3          d. Class 2 and Class 3

11. Class 3 circuit conductors cannot be taped, strapped, or attached by any means to the exterior of any conduit or other raceway as a means of support.

    a. True                    b. False

12. The input leads of a transformer or other power source supplying Class 2 and Class 3 circuits shall be permitted to be smaller than No. 14 but not smaller than No. 18 if they are not over _____ long and if they have insulation that complies with 725.49(B).

    a. 18"               b. 12'               c. 8.5"               d. 12"

13. When damage to remote-control circuits of safety control equipment would introduce a hazard, conductors of such circuits can be installed in electrical metallic tubing.

    a. True                    b. False

14. When raceway only contains Class 1 circuit conductors, the derating factors given in 310.15(B)(2)(a) do not apply if the circuit conductors carry noncontinuous loads.

    a. True                    b. False

# ● ARTICLE 727–INSTRUMENTATION TRAY CABLE: TYPE ITC

1. With the exception of No. 22 AWG conductors, the allowable ampacity of Type ITC conductors is 5 amperes.

   a. True                b. False

2. The insulation on Type ITC cable must be rated for _____.

   a. wet locations        b. 300V           c. direct sunlight usage        d. 30°C

3. Article 727 provides information pertaining construction to specification of instrumentation tray cable for application to instrumentation and control circuits operating at 125 volts or less and 5 amperes or less.

   a. True                b. False

4. Bends in Type ITC cables shall be made so as not to _____ the cable.

   a. rupture              b. minimize        c. damage                       d. stress

# ● ARTICLE 760–FIRE ALARM SYSTEMS

1.  When passing through a wall and adequate protection is not available, multiconductor NPLFA cable must be installed in either metal raceway or rigid nonmetallic conduit when the height of the wall is _____ above the floor.

    a. 10'            b. 9'            c. 7'            d. 8.5'

2.  Coaxial cables are permitted to use _____ conductivity copper-covered steel center conductor wire.

    a. 75 percent       b. 51 percent       c. 30 percent       d. 15 percent

3.  The overcurrent protection for NPLFA conductors shall not exceed _____ for No. 18 AWG conductors and _____ for No. 16 AWG conductors.

    a. 5A, 9A

    b. 10A, 7A

    c. 3A, 3.5A

    d. 4A, 12A

    e. none of the preceding

4.  Fire alarm systems include fire detection and alarm notification, guard's tour, sprinkler waterflow, and sprinkler supervisory systems.

    a. True            b. False

5.  PLFA single conductors must not be smaller than No. 18 AWG.

    a. True            b. False

6.  Transformer supplied from power-supply conductors shall be protected by an overcurrent device rated not over _____.

    a. 6A            b. 10A            c. 15A            d. 20A

7.  Fire alarm circuits are classified as _____.

    a. non-power-limited

    b. power-limited

    c. either a or b

    d. both a and b

# ● ARTICLE 770–OPTICAL FIBER CABLES AND RACEWAYS

1. The grounding conductor of optical fiber cable must be insulated and not smaller than 14 AWG copper.

   a. True              b. False

2. Optical fiber cables transmit light for _____ through an optical fiber.

   a. communication

   b. control

   c. signaling

   d. none of the preceding

   e. all of the preceding

3. One method of defining resistance to the spread of fire is for the damage not to exceed _____ when performing the vertical flame test for cables in cable trays.

   a. 3'              b. 4'11"              c. 6'7"              d. 8'6"

4. As applied for optical fiber cables, the definition of "exposed" means "on or attached to the surface or behind panels designed to allow access."

   a. True              b. False

# CHAPTER 8–EQUIPMENT FOR GENERAL USE

## ● ARTICLE 800–COMMUNICATION CIRCUITS

1. A grounding electrode shall be bonded to the metal frame of a mobile home with a copper grounding conductor not smaller than _____ when a mobile home is supplied with cord and plug.

   a. No. 6          b. No. 10          c. No. 12          d. No. 8

2. A separation of at least _____ shall be maintained between communication wires and cables on buildings and lightning conductors where practical.

   a. 11'          b. 12'          c. 6'          d. 9'

3. Where electric light, power, or NPLFA circuits exist, communication wires and cables shall be separated at least _____.

   a. 6"          b. 2"          c. 9"          d. 5"

4. "Premises" is defined as the land and buildings of a user located on the user side of the utility-user network point of _____.

   a. service          b. entrance          c. demarcation          d. connection

5. Being classified as communication cable is not required when cables are constructed of individually listed Class 2, Class 3, and communications cables under a common jacket.

   a. True          b. False

6. Communication general-purpose type cable has which of the following markings?

   a. CMX          b. CMR          c. CMG          d. CMP

7. A factory assembly of one or more insulated conductors without an overall covering is called a cable.

   a. True          b. False

8.  Not exceeding 20' in length, the primary protector grounding conductor shall be as short as practical in _____.

    a. buildings

    b. places of assembly

    c. one- and two-family dwellings

    d. utility rooms

9.  The primary protector must not be located near the point of entrance.

    a. True                    b. False

10. When communications wires and cables and electric light or power conductors are run parallel to each other in span, they must _____.

    I.  have a vertical minimum clearance of 3' when the slope of a roof is less than 4 inches in 12 inches

    II. not be attached to a cross-arm that carries electric light or power conductors

    a. only I                  b. only II                 c. both I and II
    d. either I or II          e. neither I nor II

# ● ARTICLE 810–RADIO AND TELEVISION EQUIPMENT

1.  The grounding conductor for an antenna discharge unit shall be run in as straight a line as practicable from the discharge unit to the grounding electrode.

    a. True             b. False

2.  All access doors located inside a transmitting station must be provided with interlocks that disconnect all voltages of over _____ between conductors when any door is opened.

    a. 208V         b. 250V              c. 350V                 d. 550V

3.  An outdoor antenna of a receiving station with a 47' span using a bronze conductor must not be less than _____.

    a. 19'           b. 14'                c. 17'                  d. 20'

4.  Where the maximum span between points of support is less than _____, soft- or medium-drawn copper shall be permitted for lead-in conductors.

    a. 16'      b. 24'        c. 32'        d. 40'        e. correct choice not provided

5.  An amateur station outdoor antenna conductor with a 100' span using a hard-drawn copper conductor must not be less than _____.

    a. 17'           b. 20'                c. 19'                  d. 14'

6.  A copper-clad steel grounding conductor for receiving stations shall not be smaller than _____.

    a. No. 10 AWG        b. No. 8 AWG          c. No. 17 AWG          d. No. 20 AWG

7.  Antenna conductors, where practicable, must be installed so as not to cross under open electric light or power conductors.

    a. True             b. False

# ● ARTICLE 820–COMMUNITY ANTENNA TELEVISION AND RADIO DISTRIBUTION SYSTEMS

1. Where practicable, a separation of at least _____ shall be maintained between any coaxial cable and lightning conductors.

   a. 2"            b. 1'            c. 6'            d. 10'

2. Type CATVX cable less than .375 in diameter is not permitted to be installed in _____.

   a. a condominium

   b. apartments

   c. hotels

   d. townhouses

3. Where the current supply is from a transformer and the voltage is not over _____, a coaxial cable shall be permitted to deliver low-energy power to equipment that is directly associated with the radio frequency distribution system.

   a. 50V            b. 60V            c. 70V            d. 100V

4. Selecting a grounding location to achieve the shortest practicable grounding conductor helps limit potential differences between CATV and other metallic systems.

   a. True            b. False

5. Aerial coaxial cable shall be permitted to be attached to an above-the roof raceway mast that does not enclose or support conductors of electric light or power circuits.

   a. True            b. False

6. Type CATV riser cable shall bear the cable marking _____.

   a. CATV            b. CATCR            c. CATVX            d. CATVR

# ⊙ ARTICLE 830–NETWORK-POWERED BROADBAND COMMUNICATIONS SYSTEMS

1. Over a commercial driveway, overhead spans of network-powered broadband communication cables shall not be less than _____.

    a. 9.5'              b. 11.5'              c. 15.5'              d. 17.5'

2. Type(s) _____ are cable substitutes for cable Type BLP.

    I. CL3P              II. CM              III. CMP

    a. I and II          b. II and III          c. I and III          d. none of the choices

3. A network interface unit may contain primary and secondary protectors.

    a. True              b. False

4. Where intermediate metal conduit is used to enclose network-powered broadband communication cable under a building, the conduit must be buried a minimum of _____.

    a. 6"                b. 12"                c. 4"                d. 0"

5. $VA_{max}$ represents the maximum volt-ampere output after _____ of operation, regardless of load and overcurrent protection bypassed if used.

    a. 1/4 second        b. 1 minute          c. 5 minutes          d. 1 hour

6. Steam or hot water pipes or lightning-rod conductors shall not be employed as electrodes for primary protectors.

    a. True              b. False

# CHAPTER 9–TABLES

1. The approximate diameter of a No. 12 RHW conductor is .212".
   a. True                    b. False

2. The area of a No. 1 stranded copper conductor is _____.
   a. 0.87 in²         b. 55.80 mm²         c. 0.087 in         d. .332 in²

3. $I_{max}$ limits apply after _____, where a current-limiting impedance as part of a listed product is used in combination with a stored energy source.
   a. 1 min            b. 5 sec             c. 1 hr             d. 1 min

4. The maximum length of a conduit or tubing nipple is _____.
   a. 6"               b. 10"               c. 18"              d. 24"

5. When pulling three conductors or cables into a raceway, if the ratio of the raceway to the conductor or cable is between 2.8 and 3.2, jamming can occur.
   a. True                    b. False

6. The approximate diameter of a 250 kcmil THHW Compact Copper Building Wire is _____.
   a. .670"            b. .7250"            c. .7043"           d. .6980"

7. The AC resistance of a No. 1/0 aluminum conductor installed in rigid metal conduit is _____.
   a. .002Ω            b. 20Ω               c. .20Ω             d. .02Ω

8. The allowable percent of cross-sectional area for a 4" IMC nipple is _____.
   a. 7.538 in²        b. 2.006 in²         c. 4.973 in²        d. 8.179 in²

9. Which table references power-limited fire alarm direct-current power source limitations?
   a. Table 11(A)      b. Table 12(A)       c. Table 12(B)      d. Table 11(B)

10. The resistance of a No. 3 stranded aluminum conductor is _____.
    a. .4033Ω/kFt.     b. 1.230Ω/kFt.       c. .245Ω/kFt.       d. 1.320Ω/km

11. The allowable percent of cross-sectional area of a raceway containing three conductors is 53 percent.

    a. True                 b. False

12. The circular mils area of a No. 14 stranded conductor is _____.

    a. 41100           b. 41.10           c. .4110           d. 4110

13. The actual dimensions must not be used for conductors not included in Chapter 9 such as multiconductor cables.

    a. True                 b. False

14. How many circular mils are there in a 500 kcmil conductor?

    a. 500           b. 50,000           c. 500,000           d. 5000

15. The approximate area of a 700 kcmil XHH conductor is _____ in².

    a. .856           b. .9923           c. 1.183           d. 1.065

16. Considering an 85 percent power factor, the impedance of a 750 kcmil uncoated copper conductor installed in an aluminum raceway is _____.

    a. .4Ω           b. .040Ω           c. .40Ω           d. 4Ω

17. Where bare conductors are permitted in the NEC, the dimensions for bare conductors in _____ shall be permitted.

    a. Table 4           b. Table 1           c. Table 8           d. Table 5

18. A one-shot bender is used to bend 4" IMC 90°. The bending radius of the IMC must be _____.

    a. 13"           b. 30"           c. 16"           d. 1.065"

19. The approximate diameter of a No. 12 RHW without outer covering is the same as the answer in question No. 1.

    a. True                 b. False

20. Overcurrent protection is required for an inherently limited AC power source for power-limited fire alarm circuits.

    a. True                 b. False

21. The reactance of a No. 6 conductor installed in PVC is _____.
    a. .051Ω           b. 1.61Ω                    c. .210Ω                    d. .49Ω

22. The approximate diameter of a No. 3 RHW-2 conductor is .440".
    a. True            b. False

23. The cross-sectional area of a 5" rigid metal conduit that measures 21" in length is
    _____.
    a. 12.575 in²      b. 12.692 in²               c. 12.552 in²               d. 12.127 in²

24. To the determine the maximum number of conductors of the same size allowed in raceway,
    Annex _____ must be referenced.
    a. D               b. C                        c. B                        d. A

25. The approximate area of a No. 1 THW conductor is _____.
    a. .1897 in²       b. .2006 in²                c. .1901 in²                d. .2165 in²

26. The resistance of an uncoated No. 8 stranded copper conductor is _____.
    a. .778Ω/kFt.      b. .764Ω/kFt.               c. 2.551Ω/kFt.              d. 2.506Ω/kFt.

27. The allowable percent of cross-sectional area for 1" flexible metal conduit containing four
    conductors is _____.
    a. .533 in²        b. .327 in²                 c. .283 in²                 d. .403 in²

28. The approximate area of a No. 4 XHHW Compact Copper Building Wire is _____
    in².
    a. .0730           b. .06872                   c. .0715                    d. .0746

29. The maximum overcurrent protection for a 15VAC Class 2 non-inherently limited power
    sources is _____.
    a. .5A             b. 3A                       c. 1A                       d. 5A

30. When calculating raceway fill, the actual dimensions of bonding and equipment grounding
    conductors must be used where installed.
    a. True            b. False

31. The circular mils area of a No. 10 solid conductor is _____.
    a. 10,038          b. 103,800                  c. 10,380                   d. 10.38

32. The allowable percent of cross-sectional area for a 3 1/2" PVC Schedule 80 containing 2 conductors is _____.

    a. 2.538 in²          b. 2.346 in²                    c. 2.721 in²                    d. 2.431 in²
    e. choice not available

33. A one shot bender is used to bend 2" EMT 45°. The bending radius of the EMT must be

    _____.

    a. 5 3/4"          b. 9 1/2"                    c. 7 1/4"                    d. 13"

# ANNEXES

## ● ANNEX A – PRODUCT STANDARDS

No questions.

## ● ANNEX B–APPLICATION INFORMATION FOR AMPACITY CALCULATION

1.  If thirty-five conductors were in a raceway and ten of the conductors were No. 12 THW current-carrying conductors supplying an industrial machine with a 50 percent load diversity, the ampacity of the conductors is required to be _____.

    a. 19A        b. 19.48A        c. 19.84A        d. 20A

2.  The ampacity of an underground feeder cable in electrical ducts can be taken directly from Table B.310.8.

    a. True        b. False

3.  According to Table B.310.1 a No. 10 AWG aluminum conductor must be protected by an overcurrent device not exceeding _____.

    a. 30A        b. 35A        c. 20A        d. 25A

# ANNEX C–CONDUIT AND TUBING FILL TABLES FOR CONDUCTORS AND FIXTURE WIRES OF THE SAME SIZE

1. Tables C.1–C.12 are for concentric stranded conductors only.

   a. True                 b. False

2. Compact stranding is the result of a manufacturing process where the standard conductor is compressed to the extent that the interstices (voids between strand wires) are virtually eliminated.

   a. True                 b. False

3. How many No. 8 TW conductors can be installed in a 3" EMT?

   a. 254          b. 195          c. 145          d. 81

4. Two hundred seventy-six No. 14 RHW* conductors can be installed in what size and type raceway, as listed in the Tables of Annex C?

   a. 4" IMC      b. 5" Schedule 80 PVC   c. 6" rigid steel conduit      d. no choices apply

5. When an insulation type is marked with an asterisk (*), it identifies the insulation as having an outer covering.

   a. True                 b. False

# ANSWERS TO PRACTICAL STUDY QUESTIONS

*Answers are followed by the specific article reference.*

## ⊙ ARTICLE 90–INTRODUCTION

1.  c. waive; 90.4
2.  b. soft conversion; 90.9(D), FPN No. 1
3.  c. for the practical safeguarding of persons and property; 90.1(A)
4.  b. False; 90.5(C)
5.  c. fiber optic cables and raceways; 90.2(A)

## ⊙ ARTICLE 100–DEFINITIONS

1.  b. instantaneous trip
2.  b. only II
3.  c. handhole
4.  b. outlet
5.  a. True; see definition for the term "service equipment"
6.  b. bonding
7.  b. grounded conductor
8.  c. motor control center
9.  c. service lateral
10. b. accessible, readily
11. b. False; see definition for the term "nonlinear load"
12. b. dwelling unit
13. d. all choices apply
14. b. overcurrent
15. b. general-use
16. b. False
17. c. watertight
18. b. service equipment
19. a. True; see definition for the term "wet location"

20. a. accessible (as applied to equipment)

21. c. concealed

22. b. False; see definition for the term "energized"

23. d. both a and b

24. b. False; see definition for the term "overload"

25. a. True; see definition for the term "thermal protector" (as applied to motors), FPN

26. d. utility-interactive inverter

27. b. general purpose branch circuit

28. d. all choices apply; see definition for the term "weatherproof," FPN

29. c. Branch-circuit overcurrent

30. c. ampacity

31. a. covered

32. b. interrupting rating

33. c. greatest

34. c. qualified

35. c. thermal protector

36. a. True; see definition for luminaire

37. d. all choices

38. e. I, II, and III

39. d. nonadjustable

40. a. True; see definition for attachment plug

41. d. 3

42. c. feeder

43. d. listed

44. b. rainproof

45. d. all choices apply

46. c. neutral point

47. b. service drop

48. d. continuous duty

49. d. cabinet

50. c. a shower and basin

51. a. True

52. b. fittings

53. b. damp

54. d. raintight

55. b. labeled

56. b. nominal

57. b. False

58. b. grounding electrode conductor

59. b. False; see definition for the term "guest room"

60. d. main bonding jumper

61. c. switch

62. b. False

63. b. dry

64. a. True

65. b. False; see definition for the term "ground-fault circuit interrupter," FPN


## ⊙ ARTICLE 110–REQUIREMENTS FOR ELECTRICAL INSTALLATIONS

1. e. all choices are applicable; 110.12(A)

2. b. 3.5' Table; 110.26(A)(1)

3. a. True; 110.19

4. a. True; 110.14(C)(1)(a)(1)

5. a. True; 110.31

6. b. False; 110.75(D)

7. b. 90°; 110.32

8. b. False; Table 110.20

9. b. warning signs; 110.27(C)

10. b. False; 110.2

11. d. 26 in.; 110.75(A)

12. b. 200A; 110.26(C)(2)(E), Exception

13. b. 2', 6'; 110.72, Exception

14. c. copper; 110.5

15. c. short-circuited; 110.23

16.  b. 15'; Table 110.31

17.  a. 2; 110.33(A)(1)

18.  a. True; 110.10

19.  a. True; 110.4

20.  c. orange; 110.15

21.  a. True; 110.14

22.  c. wooden plugs; 110.13(A)

23.  a. True; 110.26(F)(1)(d)

24.  c. deterioration; 110.11, FPN No. 2

25.  a. True; 110.6

26.  a. True; 110.53

27.  d. fault levels; 110.9

28.  b. snow; Table 110.20

29.  e. all of the preceding; 110.3

30.  c. Fixed; 110.79

31.  c. No. 10; 110.14(A)

32.  a. True; 110.27(A)(3)

33.  d. both a and b; 110.7

34.  b. 12.25'; Table 110.34(E)

 (Over 35kV) 125kV– 35kV = 90kV

 9.5 ft + (.37in/kV x 90kV = 33.3 in)

 9.5 ft + (33.3 in x 1ft/12in = 2.775 ft)

 9.5 ft + 2.775 ft = 12.275 ft (approximately 12.25')

## ⊙ ARTICLE 200–USE AND IDENTIFICATION OF GROUNDED CONDUCTORS

1.  b. white; 200.9

2.  b. No. 6 AWG; 200.6(A)

3.  b. False; 200.7(C)(2)

4.  f. a, b, or c; 200.10(B)(1) & (2)

5.  a. True; 200.6(B)(3)

6.  b. shell screw; 200.10(C)

7.   a. True; 200.3

8.   b. external ridge; 200.6(E)

## ⊙ ARTICLE 210–BRANCH CIRCUITS

1.   b. False; 210.11(C)(3)

2.   b. 6'; 210.52(A)(1)

3.   b. False; 210.63

4.   b. 100; 210.20(A), Exception

5.   c. 12"; 210.52(C)(1)

6.   a. True; 210.52(A)(3)

7.   a. True; 210.21(A)

8.   b. False; 210.23(A)(1)

9.   c. 12 linear feet; 210.6

10.  d. 60A; 210.3

11.  f. a and b; 210.52(E)(1)

12.  b. 2'; 210.52(A)(2)(1)

13.  a. True; 210.70(A)(1), Exception 1

14.  b. False; 210.4(C), Exceptions 1 & 2

15.  b. False; 210.8(A)(2)

16.  c. 3'; 210.52(D)

17.  e. 40A or 50A; Table 210.21(B)(3)

18.  b. 15 and 20A, 125 volt; 210.8(B)(3)

19.  b. False; 210.52(B)(1), Exception 1

20.  a. True; 210.4(A)

21.  c. 125; 210.19(A)(1)

22.  a. True; 210.6(C)(5)

23.  b. False; 210.52(H)

24.  b. No. 12; Table 210.24

25.  e. a, b, and c; 210.6(A)(1) & (2)

26.  a. 24A; Table 210.21(B)(2)

27.  c. outlets; 210.12(B)

28. e. all choices apply; 210.52(B)(1)

29. b. 208V, 240V; 210.9, Exception 1

30. d. a 4W, 3φ, 208/120 volt circuit; see definition for the term "multiwire branch circuit"

31. c. 40A; 210.19(A)(3)

## ⊙ ARTICLE 215–FEEDERS

1. c. 250.134; 215.6

2. b. False; 215.2(A)(2)

3. f. choices a–d; 215.4(A)

4. c. 1000A; 215.10

5. b. 5; 215.2(A)(3), FPN No. 2

6. b. 125; 215.2(A)(1)

7. c. 125; 215.3

## ⊙ ARTICLE 220–BRANCH-CIRCUIT, FEEDER, AND SERVICE CALCULATIONS

1. a. 1; 220.14(K)(2)

2. a. True; 220.82(A)

3. c. 4; 220.53

4. c. third largest; Table 220.103

5. f. all choices apply; 220.61(B)(1) & (2)

6. a. True; 220.18(B)

7. c. over 1.75kW; 220.55

8. a. True; 220.60

9. d. 12-month; 220.87(1)

10. b. .5; 220.5(B)

11. b. False; Table 220.55, Note 5

12. b. 30 percent; Table 220.84

13. b. 1200; 220.14(F)

14. d. 5000 watts or the nameplate rating, whichever is larger; 220.54

15. b. False; 220.42

16. a. True; 220.51

17. c. 65 Table; 220.56

18. c. 2 ft; 220.43(B)

19. a. 1/8 hp; 220.18(A)

20. b. False; 220.12

21. d. two-wire; 220.52(A) & (B)

22. b. 50; Table 220.44

23. b. 180; 220.14(I)

## ⊙ ARTICLE 225–OUTSIDE BRANCH CIRCUITS AND FEEDERS

1. b. 13.5'; Table 225.61

2. a. True; 225.35

3. b. False; 225.26

4. b. 12'; 225.18(2)

5. c. 2000A; 225.30(C)

6. a. 60A; 225.39(D)

7. b. 4"; 225.14(C)

8. c. 36"; 225.19(D)(1), Exception

9. b. No. 12; 225.6(B)

10. b. raintight, drain; 225.22

11. a. True; 225.4

## ⊙ ARTICLE 230–SERVICES

1. a. True; 230.33

2. a. True; 230.74

3. d. 35,000 volts; 230.212

4. a. True; 230.82(6)

5. d. 2.5"; Table 230.51(C)

6. g. III and IV; 230.23(B)

7. a. True; 230.54(F)

8.  d. 1 second; 230.95(A)

9.  d. 6; 230.71(A)

10. a. True; 230.2

11. c. 100A; 230.79(C)

12. a. True; 230.75

13. c. isolating; 230.204(A)

14. c. bathrooms; 230.70(A)(2)

15. c. bare; 230.22, Exception

16. c. circuit breaker; 230.90(B)

17. d. 300V; 230.24(A), Exception 2

18. e. all of the preceding; 230.50(B)(1)

19. c. 10'; 230.26

20. b. False; 230.3

21. a. True; 230.41, Exception (5)

22. a. True; 230.40, Exception 4

23. b. grounded; 230.95

24. b. 3'; 230.9(A)

25. a. True; 230.90(A)

26. c. both a and b; 230.54(A) & (B)

27. d. only a and b; 230.81

28. c. 18'; 230.24(B)(4)

29. b. 125; 230.42(A)(1)

30. e. I, II, and III are applicable; 230.7, Exceptions No. 1 & 2


## ⊙ ARTICLE 240–OVERCURRENT PROTECTION

1.  b. False; 240.10

2.  a. it will open both ungrounded and grounded conductors at the same time; 240.22(1)

3.  d. 30; 240.51(A)

4.  d. overcurrent trip unit; 240.15(A)

5.  b. False; 240.21(B)

6.  b. False; 240.50(A)(2)

7. b. 84A; 240.21(B)(2)(1)

8. c. three; 240.100(A)(1)

9. c. load; 240.50(E)

10. a. True; 240.1, FPN

11. d. 250A; 240.4(B)(2)

12. b. False; 240.23

13. c. S; 240.52

14. a. True; 240.61

15. c. 25A; 240.4(D)(6)

16. c. both a and b; 240.30(A)(1)

17. f. a, b, or c; 240.5(B)(2)

18. c. ampere; 240.50(B)

19. a. True; 240.85

20. c. tap; 240.2, definition

21. b. False 240.15(B)

22. c. readily accessible; 240.24(A)

23. c. 125; 240.53(A)

24. a. True; 240.24(E) see definition for the term "bathroom"

25. a. True; 240.21(B)(1)(2)

26. c. 5000; 240.83(C)

27. b. No; 240.4(F) & 240.21(C)(1)

28. d. 100; 240.21(B)(4)(2)

29. e. a and c; 240.8

30. c. 800; 240.4(B)(3)

31. b. multiwire; 240.15(B)(1)

32. b. False; 240.24(D)

33. f. all choices apply; 240.60(C)

34. a. True; 240.21

35. f. b and c; 240.81

36. c. both choices applicable; 240.33

37. b. grouped; 240.21(C)(3)(3)

38. b. 5A; 240.6(A)

1.  d. grounding electrode conductor; 250.24(C)(1)

2.  c. 10 feet; 250.52(A)(1)

3.  c. No. 6; 250.64(B)

4.  b. fault; 250.96(A)

5.  d. neither a nor b; Table 250.122

6.  d. 100 volts; 250.188(C)

7.  g. a, b, and c; 250.53(G)

8.  c. either a or b; 250.36(B)

9.  d. 6; 250.56

10. a. True; 250.2, definition

11. d. neither a nor b; Table 250.66

12. d. 20 feet; 250.52(A)(3)

13. a. True; 250.64(C)(1)

14. c. No. 14; 250.102(D)

15. a. multiconductor cable having three insulated conductors with an equipment ground; 250.134(B), 250.140

16. g. I, II, and III; 250.34(A)(1) & (2)

17. c. 10; 250.178

18. d. not be grounded; 250.22(4) and (5)

19. a. True; 250.94, FPN 1

20. c. voltage; 250.4(A)(1)

21. c. circular mil area; 250.24(C)(2)

22. d. either a or b; 250.166(A)

23. f. c or d; 250.66(A)

24. b. 6; 250.102(E)

25. a. True; 250.140, Exception

26. h. all choices apply; 250.118(1)

27. a. True; 250.20(D), FPN 1

28. b. 25Ω; 250.56

29. a. True; 250.24(A)(5)

30. b. False; 250.4(A)(5)

31. a. True; 250.26(3)

32. c. 3/4; 250.52(A)(5)

33. d. 1100, 1750; 250.24(C)(1)

34. b. equipment grounding; 250.104(A)(2)

35. a. True; 250.146(D)

36. b. False; 250.122(B)

37. c. Bonding jumpers; 250.92(B)

38. d. 1300'; 250.184(C)

39. a. True; 250.36(F)

40. e. circuit conductors, surge-protective devices; 250.6(A)

41. d. all choices apply; 250.28(A)

42. a. True; 250.52(A)(7)

43. b. False; 250.68(A)

44. e. Switchboard Frames, 2-wire DC effectively insulated from ground; 250.112(A)

45. b. 50; 250.162(A)

46. b. 2.5'; 250.53(F)

47. b. False; 250.119

48. d. 6; 250.119(A)(1)

49. a. True; 250.86, Exception 3

50. a. True; 250.12

51. a. True; 250.30(A)(7)(2), Exception 2

52. b. False; 250.52(B)

53. c. 1; 250.70

54. a. True; 250.114(3) & (4)

55. c. 2 AWG; 250.52(A)(4)

56. d. 300 volts; 250.170

57. a. True; 250.36(A)

58. d. 6; 250.119(A)(1)

59. b. False; 250.92(B)

60. b. 4W/3φ/ delta; 250.20(B)(3)

## ⊙ ARTICLE 280–SURGE ARRESTERS, OVER 1kV

1. f. a, b, or c; 280.12
2. c. 60Hz; 280.24(B)(1)
3. c. surge arrester; 280.2
4. c. 6; 280.23
5. b. False; 280.11
6. b. 125; 280.4(B)


## ⊙ ARTICLE 285–SURGE-PROTECTIVE DEVICES (SPDs), 1kV OR LESS

1. a. first; 285.24(C)
2. b. ungrounded; 285.4
3. a. True; 285.27
4. b. False; 285.3(1)


## ⊙ ARTICLE 300–WIRING METHODS

1. b. 1/16"; 300.4(B)(2)
2. c. thermal; 300.7(B)
3. d. 36"; Table 300.50
4. b. warning ribbon; 300.5(D)(3)
5. c. mineral-insulated; 300.15(D)
6. a. True; 300.5(J), FPN
7. a. joined; 300.10
8. a. No. 4 AWG; 300.4(G)
9. d. 1/4"; 300.6(D)
10. a. True; 300.1(A)
11. c. 1 1/4"; 300.4(A)(1)
12. a. True; 300.8
13. a. True; 300.5(E)
14. e. all of the preceding; 300.3(B)

15. d. 6"; 300.14

16. d. 8'; 300.5(D)(1)

17. b. induction; 300.20(A)

18. c. insulation; 300.3(C)(1)

19. b. False; 300.4(A)(2), Exception 1

20. From Table 300.1(C):

| Metric Designator | Trade Sizes |
|---|---|
| 12 | 3/8 |
| 21 | 3/4 |
| 35 | 1 1/4 |
| 53 | 2 |
| 91 | 3 1/2 |
| 103 | 4 |
| 155 | 6 |

21. c. cover; Table 300.5, Note 1

22. b. False; 300.3(C)(2)

23. c. 12; 300.34

24. d. 24"; Table 300.5

25. a. True; 300.3(C)(1)

26. a. 1"; 300.39

27. c. bushing; 300.5(H)

28. d. multiwire; 300.13(B)

29. d. 135'; Table 300.19(A)

30. c. 600; 300.2(A)

## ⊙ ARTICLE 310–CONDUCTORS FOR GENERAL WIRING

1. a. True; 310.12(C)

2. a. True; Table 310.13(A)

3. a. True; Table 310.15(B)(2)(a), FPN No. 1

4. c. 135A; Table 310.16

5. a. True; 310.15(B)(4)(a)

6.  d. both a and b; 310.6

7.  c. machine tool wiring; Table 310.13(A)

8.  a. True; see Correction Factors, Tables 310.16 & .17

9.  a. True; 310.15(B)(3)

10. a. True; 310.2(A)

11. a. True; Tables 310.16–20

12. d. 24"; 310.15(B)(2), Exception 3

13. d. 250; Table 310.15(B)(6)

14. d. No. 4/0; Table 310.13(A)

15. f. all of the preceding; 310.4(B)

16. b. lowest; 310.15(A)(2)

17. h. a, b, c, and d; 310.8(B)

18. a. True; 310.8(C)

19. e. not applicable; (40°F) Table 310.15(B)(2)(c)

20. b. 8; 310.3

21. b. M; 310.11(C)

22. c. can remain as is; Heading, Table 310.16

23. b. 0

24. d. maximum; 310.10, FPN 1

25. b. 14, 12; Table 310.5

26. b. 138A; Table 310.20

27. b. False; 310.15(B)(2)(c)

28. e. all of the preceding; 310.11(A)

29. b. 70; Table 310.15(B)(2)(a)

30. c. No. 1/0 AWG; 310.4

## ⊙ ARTICLE 312–CABINETS, CUTOUT BOXES, AND METER SOCKET ENCLOSURES

1. a. True; Table 312.6(B), Note 2
2. a. 30A; 312.11(B)
3. c. 4; 312.6(C)
4. c. 1"; 312.11(A)(2)
5. a. True; 312.2
6. c. 0.053"; 312.10(B)
7. d. all of the preceding; 312.8
8. b. 8"; Table 312.6(A)


## ⊙ ARTICLE 314–OUTLET, DEVICE, PULL, AND JUNCTION BOXES; CONDUIT BODIES; FITTINGS; AND HANDHOLE ENCLOSURES

1. b. 1; 314.16(B)(3)
2. c. 50 lbs; 314.27(B)
3. c. 1/2; 314.72(E)
4. d. I, II, and III; 314.16(A)
5. c. 1/4"; 314.20
6. b. False; 314.28(A)(2), Exception
7. c. 4" x 1 1/4"; Table 314.16(A)
8. e. not applicable; 314.16(B)(4)
9. e. all of the preceding; 314.25
10. b. False; 314.2
11. d. 1" x 2"; 314.23(B)(2)
12. b. False; 314.27(D)
13. a. True; 314.16(C)(1)
14. d. none of the preceding; 314.40(B)
15. a. True; 314.16(A)(2)
16. b. False; 314.16(B)(1)
17. a. True; 314.16(B)(1)
18. d. 8; 314.28(A)(1)

19. d. 3 in³; Table 314.16(B)

20. d. 6; 314.5

21. e. answer not provided;

| | |
|---|---|
| 2 cable clamps (#12) | 2.25 in³ x 1 = 2.25 in³ |
| 2 #14 | 2.00 in³ x 2 = 4.00 in³ |
| 3 #12 | 2.25 in³ x 3 = 6.75 in³ |
| 1 #12 ground | 2.25 in³ x 1 = 2.25 in³ |
| 2 switches (#12) | 2.25 in³ x 4 = 9.00 in³ |
| 1 receptacle (#14) | 2.00 in³ x 2 = <u>4.00 in³</u> |
| | 28.25 in³ |

A box having a minimum cubic inch capacity of 28.25 in³ is required.

22. a. 15/16"; 314.24(B)

23. c. 4; Table 314.16(A)

24. b. 8"; 314.17(C), Exception

25. b. 36; 314.71(B)

26. a. True; 314.15

27. a. True; 314.23(H)(1)

28. c. 13.5 in³; Table 314.16(A)


## ⊙ ARTICLE 320–ARMORED CABLE: TYPE AC

1. b. internal bonding; 320.100

2. b. False; 320.30(B)

3. a. I and II; 320.12

4. c. 5; 320.24

5. c. 60°C; 320.80(A)

6. c. 6'; 320.23(A)

7. a. True; 320.2, definition

## ⊙ ARTICLE 322–FLAT CABLE ASSEMBLIES: TYPE FC

1. d. 10; 322.104
2. a. True; 322.40(B)
3. e. none of the preceding; 322.120(C)
4. c. 30A; 322.10(1)
5. b. 15A; 322.56(B)

## ⊙ ARTICLE 324–FLAT CONDUCTOR CABLE: TYPE FCC

1. c. equipment grounding; 324.100(A)
2. c. 30; 324.10(B)(2)
3. b. False; 324.40(A)
4. d. carpet squares; 324.2, definition
5. d. all of the preceding; 324.120(A)
6. c. three; 324.2, definition
7. b. 150; 324.10(B)(1)
8. c. .09"; 324.10(G)

## ⊙ ARTICLE 326–INTEGRATED GAS SPACER CABLE: TYPE IGS

1. d. 491A; Table 326.80
2. a. True; 326.26
3. c. 3.710"; Table 326.116
4. b. False; 326.12
5. b. 20; 326.112
6. d. 35"; Table 326.24

## ⊙ ARTICLE 328–MEDIUM VOLTAGE CABLE: TYPE MV

1. d. 2001; 328.2, definition
2. b. False; 328.10

## ⊙ ARTICLE 330–METAL-CLAD CABLE: TYPE MC

1.  d. 402.5; 330.80
2.  c. No. 10; 330.30(B)
3.  b. 10; 330.24(A)(1)
4.  b. False; 330.10(A)(5)
5.  b. 18, 12; 330.104
6.  c. minimize; 330.10(A)(12)
7.  c. metal-clad; 330.2, definition

## ⊙ ARTICLE 332–MINERAL-INSULATED, METAL-SHEATHED CABLE: TYPE MI

1.  d. all of the preceding; 332.104
2.  b. 10; 332.24(2)
3.  c. equipment grounding; 332.108
4.  b. False; 332.40(B)
5.  b. False; 332.10(7)
6.  b. 6; 332.30

## ⊙ ARTICLE 334–NONMETALLIC-SHEATHED CABLE: TYPES NM, NMC, AND NMS

1.  b. 60°C; 334.80
2.  b. False; 334.24
3.  a. True; 334.2, definitions
4.  d. No. 2 AWG; 334.104
5.  a. True; 334.40(B)
6.  c. corrosion resistant; 334.2, definitions
7.  b. False; 334.108
8.  a. True; 334.30(B)(1)
9.  e. b and c; 334.15(C)
10. d. I, II, and III; 334.10(1) & (2)
11. d. metal or Schedule 80 nonmetallic raceway extending at least 6" above the floor; 334.15(B)
12. c. 12"; 334.30

## ⊙ ARTICLE 336–POWER AND CONTROL TRAY CABLE: TYPE TC

1.  b. False; 336.104
2.  c. 6; 336.24(3)
3.  e. all of the preceding; 336.10

## ⊙ ARTICLE 338–SERVICE-ENTRANCE CABLE: TYPES SE AND USE

1.  a. True; 338.2 definitions
2.  b. 1; 338.100
3.  c. equipment grounding; 338.10(B)(2)

## ⊙ ARTICLE 340–UNDERGROUND FEEDER AND BRANCH-CIRCUIT CABLE: TYPE UF

1.  d. 4/0; 340.104
2.  a. 60°C; 340.80
3.  b. False; 340.12
4.  b. False; 340.116

## ⊙ ARTICLE 342–INTERMEDIATE METAL CONDUIT: TYPE IMC

1.  b. 10; 342.30(B)(1)
2.  c. 4"; 342.20(B)
3.  b. False; 342.14
4.  a. True; 342.60
5.  c. 3/4" taper/ft; 342.28
6.  b. False; 342.6
7.  b. 360°; 342.26

# ⊙ Article 344–Rigid Metal Conduit: Type RMC

1.  a. True; 344.10(D)
2.  a. True; 344.42(B)
3.  c. 20'; Table 344.30(B)(2)
4.  d. 6"; 344.20(B)

# ⊙ Article 348–Flexible Metal Conduit: Type FMC

1.  d. 3; Table 348.22
2.  b. False; 348.42
3.  d. flexible metal conduit; 348.2
4.  b. False; 348.12(6)

# ⊙ Article 350–Liquidtight Flexible Metal Conduit: Type LFMC

1.  a. True; 350.30(A), Exception 1
2.  a. True; 350.12
3.  a. True; 350.120

# ⊙ Article 352–Rigid Polyvinyl Chloride Conduit: Type PVC

1.  c. trimmed; 352.28
2.  a. 2.03; Table 352.44
3.  c. 6'; Table 352.30
4.  a. True; 352.10(D)
5.  d. 6"; 352.20(B)
6.  a. True; 352.60, Exception 2
7.  a. 50°C; 352.12(D)

## ⊙ ARTICLE 353–HIGH DENSITY POLYETHYLENE CONDUIT: TYPE HDPE CONDUIT

1. b. electrical; 353.2, definition
2. b. False; 353.60
3. a. 1/2", 6"; 353.20(A) & (B)

## ⊙ ARTICLE 354–NONMETALLIC UNDERGROUND CONDUIT WITH CONDUCTORS: TYPE NUCC

1. a. True; 354.12(3)
2. b. nonmetallic underground conduit with conductors; 354.2, definition
3. a. both ends; 354.120
4. e. choice not provided; Table 354.24

## ⊙ ARTICLE 355–REINFORCED THERMOSETTING RESIN CONDUIT: TYPE RTRC (NEW)

1. a. True; 355.120
2. b. False; 355.12(D)

## ⊙ ARTICLE 356–LIQUIDTIGHT FLEXIBLE NONMETALLIC CONDUIT: TYPE LFNC

1. b. False; 356.30(3)
2. c. 1; 356.10(6)
3. c. LFNC; 356.2, FPN
4. a. LFNC-C; 356.2(3), definition

## ⊙ ARTICLE 358–ELECTRICAL METALLIC TUBING: TYPE EMT

1. b. False; 358.28(B)
2. d. all of the preceding; 358.10(B)
3. c. listed; 358.42
4. c. 10; 358.30(A)
5. a. True; 358.2, definition

## ⊙ ARTICLE 360–FLEXIBLE METALLIC TUBING: TYPE FMT

1. c. 17.5"; Table 360.24(A)
2. a. True; 360.12
3. c. FMT; 360.2, definition

## ⊙ ARTICLE 362–ELECTRICAL NONMETALLIC TUBING: TYPE ENT

1. c. 2; 362.20(B)
2. a. True; 362.60
3. a. True; 362.12
4. a. True; 362.2, definition
5. a. True; 362.46
6. c. 6'; 362.30(A), Exception 1
7. c. three; 362.10(2)

## ⊙ ARTICLE 366–AUXILIARY GUTTERS

1. a. True; 366.10(B)(1), FPN
2. b. busbars; 366.100(E)
3. c. 30; 366.12(2)
4. a. True; 366.120(A)(2)
5. d. 3'; 366.30(B)
6. d. 75; 366.56(A)

7.  b. False; 366.12(1)

8.  a. 30; 366.23(A)


## ⊙ ARTICLE 368–BUSWAYS

1.  d. all of the preceding; 368.17(B), Exception

2.  a. True; 368.10(C)(1)

3.  a. I; 368.10(B)(1)

4.  a. True; 368.17(C)

5.  b. False; 368.12(E)

6.  a. True; 368.234(A), Exception

7.  d. busway; 368.2, definitions

8.  e. only b and c; 368.320

9.  b. 5'; 368.30

10. a. True; 368.258

11. c. current rating; 368.17(A)


## ⊙ ARTICLE 370–CABLEBUS

1.  c. 310.69 and .70; 370.4(B)

2.  c. fault; 370.2, definition

3.  c. 1/0; 370.4(C)

4.  a. True; 370.3

5.  b. False; 370.4(D)


## ⊙ ARTICLE 372–CELLULAR CONCRETE FLOOR RACEWAYS

1.  a. True; 372.12

2.  d. 90° angle; 372.5

3.  b. 40; 372.11

4.  a. True; 372.2, definition

## ⊙ ARTICLE 374–CELLULAR METAL FLOOR RACEWAYS

1.   a. True; 374.8

2.   b. shall be removed from the raceway; 374.7

3.   a. True; 374.3(1)

## ⊙ ARTICLE 376–METAL WIREWAYS

1.   b. 4; 376.23(B)

2.   a. True; 376.58

3.   b. metal wireway; 376.2, definition

4.   c. 15; 376.30(B)

5.   b. listed; 376.10(3)

## ⊙ ARTICLE 378–NONMETALLIC WIREWAYS

1.   a. 20; 378.22

2.   d. 75; 378.56

3.   b. flame retardant; 378.2, definition

4.   a. True; 378.10(4)

## ⊙ ARTICLE 380–MULTIOUTLET ASSEMBLY

1.   a. True; 380.3

2.   b. False; 380.2(B)(3)

## ⊙ ARTICLE 382–NONMETALLIC EXTENSIONS

1.   a. True; 382.56

2.   d. 2; 382.15(A)

3.   a. True; 382.10(A)

# ⊙ ARTICLE 384–STRUT-TYPE CHANNEL RACEWAY

1.  a. True; 384.100(C)
2.  c. 75; 384.56
3.  a. True; 384.10(4)

# ⊙ ARTICLE 386–SURFACE METAL RACEWAYS

1.  d. identified; 386.100
2.  b. Class I, Division 2; 386.10(2)

# ⊙ ARTICLE 388–SURFACE NONMETALLIC RACEWAYS

1.  b. False; 388.22
2.  b. False; 388.70

# ⊙ ARTICLE 390–UNDERFLOOR RACEWAYS

1.  a. True; 390.3(B)
2.  c. 40; 390.5
3.  d. both a and b; 390.2(A)
4.  c. 4, 3/4; 390.3(A)
5.  c. concrete; 390.13

# ⊙ ARTICLE 392–CABLE TRAYS

1.  d. 2000A; Table 392.7(B)
2.  a. True; 392.5(F)
3.  a. 9"; 392.3(B)(1)(a)
4.  b. cable tray system; 392.2, definition
5.  a. True; 392.8(E)

6. d. 6'; 392.6(A)

7. c. sunlight resistant; 392.3

## ● ARTICLE 394–CONCEALED KNOB-AND-TUBE WIRING

1. a. True; 394.56

2. b. equivalent; 394.30(B)

3. a. 3; 394.19(A)

## ● ARTICLE 396–MESSENGER-SUPPORTED WIRING

1. a. True; 396.12

2. c. messenger-supported wiring; 396.2(3)

## ● ARTICLE 398–OPEN WIRING ON INSULATORS

1. d. 12"; 398.30(A)(2)

2. a. True; 398.10(1)

3. a. True; 398.30(D)

4. a. 4.5; 398.30(B)

5. d. 2"; 398.19

## ● ARTICLE 400–FLEXIBLE CORDS AND CABLES

1. a. True; 400.10, FPN

2. b. 50; Table 400.4, Note 4

3. a. 12; 400.31(A)

4. d. 6; Table 400.4

5. c. 101A; Table 400.5(B), Note 3

6. e. all of the preceding; Table 400.4

7. b. stresses; 400.31(B)

8. b. Type SPT-3; Table 400.4

9. c. light blue; 400.22(C)

10. b. False; Table 400.4

11. c. 45; 400.21, Exception

12. b. False; 400.8(2)

13. c. extra hard; Table 400.4

14. a. 5A; Table 400.5(A)


## ⊙ ARTICLE 402–FIXTURE WIRES

1. b. pressure; 402.3, FPN

2. b. 23; Table 402.5

3. b. 5.5; Table 402.3

4. c. branch-circuit; 402.11

5. c. 194; Table 402.3

6. d. 600; 402.3

7. a. True; 402.6

8. c. 300; Table 402.3


## ⊙ ARTICLE 404–SWITCHES

1. d. answer not available; 404.14(A)(3)

2. c. 300; 404.8(B)

3. d. 200A; 404.16

4. c. either a or b; 404.6(B)

5. a. .010"; 404.9(C)

6. b. at the same time; 404.2(B), Exception

7. a. True; 404.17

8. b. 6'7"; 404.8(A)

9. a. and c.; 404.14(D)(1) & (2)

10. b. ungrounded; 404.2(A)

11. b. 50; 404.14(B)(2)

12. b. False; 404.13(B)

13. c. either a or b; 404.4

14. e. all of the preceding; 404.9(B)(1) & (2)


## ⊙ ARTICLE 406–RECEPTACLES, CORDS CONNECTORS, AND ATTACHMENT PLUGS (CAPS)

1. a. True; 406.3(B), Exception 1

2. d. all of the preceding; 406.9(B)(1)(2) & (3)

3. d. orange triangle; 406.2(D)

4. c. either a or b; 406.2(B)

5. a. weatherproof; 406.8(B)(1)

6. f. all of the preceding; 406.3(D)(3)

7. b. False; 406.2(C)


## ⊙ ARTICLE 408–SWITCHBOARDS AND PANELBOARDS

1. b. False; 408.36(B), Exception

2. b. 10; Table 408.5

3. c. B; 408.3(E)

4. a. 3; 408.18(A)

5. b. False; 408.36, Exception 3

6. d. all of the preceding; 408.3(E)

7. a. True; 408.36

8. a. True 408.19

9. b. False; 408.41

10. a. True; 408.1(2)

11. a. True; 408.36(C)

12. c. 30A, 200A; 408.36(A)

13. a. 1/2"; Table 408.56

# ⊙ ARTICLE 409–INDUSTRIAL CONTROL PANELS

1. b. False; 409.108
2. b. equipment-grounding conductor; 409.60
3. b. 125; 409.20

# ⊙ ARTICLE 410–LUMINAIRES, LAMPHOLDERS, AND LAMPS

1. a. 20; 410.30(B)(1), Exception 2
2. d. .04"; 410.78(A)
3. b. False; 410.130(B)
4. a. True; 410.2, definitions
5. c. weatherproof; 410.96
6. b. branch-circuit overcurrent device; 410.59(A)
7. d. 12; 410.55(A)
8. b. False; 410.21
9. c. 1/2"; 410.86
10. b. False; 410.5, Exception
11. b. grounded; 410.138
12. b. False; 410.78(B)
13. c. thermal; 410.130(E)(3)
14. d. none of the preceding; 410.103
15. b. 150; 410.143(C)
16. f. a, b, c, and d; 410.16(A)(1) & (2)
17. d. 50; 410.62(C)(2)
18. d. 302°F; 410.115(B)
19. c. closet storage space; 410.2, definitions
20. b. Wet; 410.10(A)
21. b. connected to the screw shell; 410.50
22. a. True; 410.1
23. d. 6 lbs; 410.30(A)
24. d. all of the preceding; 410.140(B)

25.  b. 8; 410.12

26.  b. False; 410.54(C)

27.  b. heat; 410.104(A)

28.  a. 3"; 410.68

29.  a. True; 410.2, definitions

30.  c. 3, 8; 410.10(D)

31.  c. 18; 410.54(B)

32.  e. none of the preceding; 410.136(B), FPN

33.  a. True; 410.36(C)

34.  b. False; 410.93

35.  a. Stranded; 410.56(E)

36.  b. 20; 410.153

37.  a. 6; 410.16(C)(4)

38.  b. 50; 410.82(B)(5)

39.  a. True; 410.74(A)

40.  b. False; 410.11

41.  a. 1'; 410.117(C)


## ⊙ ARTICLE 411–LIGHTING SYSTEMS OPERATING AT 30 VOLTS OR LESS

1.  c. 20; 411.6

2.  a. 7'; 411.5(C)

3.  a. True; 411.4(B)


## ⊙ ARTICLE 422–APPLIANCES

1.  c. either a or b; 422.31(B)

2.  d. none of the preceding; 422.11(B)

3.  a. True; 422.18

4.  d. 120; 422.11(F)(3)

5.  a. True; 422.33(B)

6.  b. 300; 422.48(B)

7.  b. 150; 422.11(E)(3)

8.  e. either a, b or c; 422.61

9.  a. 18", 36"; 422.16(B)(1)(2)

10. a. 125; 422.10(A)

11 . a. True; 422.12

12. c. electrocution; 422.41

13. c. ground-fault circuit-interrupter; 422.49

14. d. none of the preceding; 422.13

15. a. temperature-limiting; 422.46

16. d. 4; 422.16(B)(2)(2)


## ⊙ ARTICLE 424–FIXED ELECTRIC SPACE-HEATING EQUIPMENT

1.  c. 2; 424.36

2.  a. True; 424.99(C)(5)

3.  b. energized; 424.63

4.  a. 2"; 424.39

5.  a. total load; 424.82

6.  d. 7; 424.34

7.  d. both a and b; 424.44(G)

8.  b. 60A; 424.22(B)

9.  b. heating panel; 424.91, definitions

10. e. all choices apply; 424.3(A)

11. d. all choices apply; 424.19

12. d. 33 watt/ft$^2$; 424.98(A)

13. a. True; 424.59, FPN

14. b. False; 424.22(E)

15. d. 16.5; 424.44(A)

16. b. receptacle; 424.9, FPN

17. a. True; 424.19(B)(1)

18. c. 120 volt

    d. 208 volt

a. 240 volt

e. 277 volt

b. 480 volt; 424.35

19. a. True; 424.41(B)

20. d. 125; 424.3(B), see 210.19(A)(1)


## ⊙ ARTICLE 426–FIXED OUTDOOR ELECTRIC DEICING AND SNOW-MELTING EQUIPMENT

1. a. 1 to 6; 426.22(B)

2. a. True; 426.2, FPN

3. c. dual winding; 426.31

4. a. True; 426.50(A)

5. d. 120 watts/ft$^2$; 426.20(A)

6. b. 80; 426.34

7. b. skin-effect heating system; 426.2, definition

8. a. True; 426.24(B)

9. c. 125; 426.4, see 210.19(A)(1)

10. g. a, b and c; 426.20(E)

11. b. False; 426.12


## ⊙ ARTICLE 427–FIXED ELECTRIC HEATING EQUIPMENT FOR PIPELINES AND VESSELS

1. a. True; 427.18(A)

2. b. watertight; 427.46

3. d. 20'; 427.13

4. a. True; 427.2, definitions

5. d. not applicable; 427.30

6. a. 125; 427.4, see 210.19(A)(1)

1. c. 3/8"; Table 430.12(C)(1)
2. d. not less than 125 percent of the full-load current of the highest rated motor plus the sum of the full-load current ratings of all other motors in the group; 430.24
3. a. 15; 430.42(C)
4. c. 500; 430.72(C)(4)
5. a. True; 430.104
6. d. 1.41; Table 430.249
7. d. 85; Table 430.22(E)
8. c. 6; 430.53(A)(1)
9. b. one; Table 430.37
10. c. standard part; 430.4
11. a. True; 430.82(A)
12. b. False; 430.28(2)
13. a. True; 430.43
14. d. 90; Table 430.72(B)
15. b. False; 430.109(G)
16. a. 110; Table 430.250
17. a. True; 430.7(C)
18. a. True; 430.88, Exception
19. a. True; 430.72(C), Exception
20. b. 150; 430.4, Exception
21. a. True; 430.14(B), Exception
22. a. True; 430.124(A)
23. d. 150; Table 430.52
24. b. 12; Table 430.10(B), footnote, see Table 312.6(B)
25. c. 115; 430.110(A)
26. a. True; 430.6(A)(1)
27. c. 80 percent; 430.83(C)(2)
28. c. motor control circuit; 430.2, definition
29. d. both a and b; 430.2, definition
30. b. current; 430.6(A)(1)

31. b. full-load current; 430.17

32. a. True; 430.31, FPN

33. c. 10; 430.245(B)

34. a. 1/3; 430.81(B)

35. b. False; 430.111(B)(2)

36. b. 7; 430.9(C)

37. a. 115; 430.226

38. c. 3-wire, three phase; 430.36

39. c. 65; Table 430.23(C)

40. b. False; 430.6(A)(2)

41. c. motor's full-load current; 430.22(A)

42. b. II and III; 430.32(A)(1)

43. a. True; 430.52(C)(3), FPN

44. d. 110; Table 430.12(B)

45. a. 6'; 430.223

46. b. False; 430.83(A)(1)

47. d. 400; 430.35(A)

48. a. True; Table 430.22(E), Footnote

49. a. True; 430.243

50. b. arc welders; 430.7(A)(7)

51. a. 50; 430.22(D)

52. b. 140; 430.32(A)(2)

53. b. nameplate; 430.52(D)

54. d. all of the preceding; 430.82(B)

55. c. 15, 30; Table 430.29

56. a. True; 430.126(A), FPN

57. b. False; 430.62(A)

58. c. horizontally; 430.97, Exception

59. a. 225 percent; 430.52(C)(1), Exception 2(b)

60. c. 4.5 to 4.99; Table 430.7(B)

61. a. 14; 430.22(F)

62. a. True; 430.32(C)

# ⊙ Article 440–Air-Conditioning and Refrigerating Equipment

1. b. False; 440.55(B)
2. e. the larger of a or b; 440.12(A)
3. c. 12"; 440.65
4. e. only a and c; 440.4(A)
5. a. 25; 440.33
6. h. a, b, and c; 440.2, definitions
7. c. 175; 440.22(A)
8. b. False; 440.6(A), Exception 1
9. d. none of the preceding; 440.64
10. e. a or c; 440.2, definitions
11. b. 50; 440.62(C)
12. d. 140; 440.52(A)(1)
13. g. a, b, or c; 440.3(B)


# ⊙ Article 445–Generators

1. c. 65; 445.12(C)
2. a. True; 445.12(B)
3. c. 115; 445.13


# ⊙ Article 450–Transformers and Transformer Vaults (Including Secondary Ties)

1. c. both a and b; 450.24
2. b. noncombustible moisture-resistant; 450.8(B)
3. b. a drain; 450.46
4. b. False; 450.5, FPN
5. b. False; 450.42
6. b. False; 450.45(E)
7. b. Class 155; 450.22

8.  a. True; 450.3, FPN 2

9.  b. 1 hour; 450.21(B)

10. d. 1; 450.42, Exception

11. b. False; 450.6(B)

12. d. all of the preceding; 450.2, definition

13. b. 3; 450.42

14. a. 42; 450.5(B)(2)(b)

15. d. 35kV; 450.21(C)

16. b. 600; 450.6

17. d. 400; Table 450.3(A)


## ⊙ ARTICLE 455–PHASE CONVERTERS

1.  a. True; 455.7(A) & (B)

2.  a. True; 455.23

3.  c. 250 percent; 455.6(A)(2)

4.  b. static phase converter; 455.2, definition

5.  b. multiplied by; 455.8(D)

6.  d. all of the preceding; 455.8(B)

7.  b. variable; 455.6(A)(1)


## ⊙ ARTICLE 460–CAPACITORS

1.  a. True; 460.8(B)

2.  c. 1 minute; 460.6(A)

3.  c. 135; 460.8(A)

4.  a. 3; 460.2(A)

5.  b. 50; 460.28(A)

6.  c. improved power factor; 460.9

## ● ARTICLE 470–RESISTORS AND REACTORS

1.  d. 12; 470.18(C)

2.  c. 90°C 470.4

3.  a. True; 470.2

## ● ARTICLE 480–STORAGE BATTERIES

1.  d. all choices apply; 480.4

2.  d. pressure relief vent; 480.10(B)

3.  b. False; 480.2, definition

4.  c. storage; 480.2, definition

5.  d. correct choice not provided; 480.6(C)

6.  b. False; 480.6(B)

## ● ARTICLE 490–EQUIPMENT, OVER 600 VOLTS, NOMINAL

1.  c. 15"; Table 490.24

2.  d. all of the preceding; 490.21(A)(1)

3.  a. True; 490.44(B)

4.  d. 25; 490.72(D)

5.  c. 5; 490.21(D)(7)

6.  c. 3-phase, 4-wire, solidly grounded wye system; 490.71

7.  d. 10; 490.3

8.  a. 66"; 490.41(A), Exception

9.  b. False; 490.21(B)(1)

10. d. either I, II, or III; 490.23

11. c. conspicuous sign; 490.21(B)(7), Exception

## ⊙ ARTICLE 500–HAZARDOUS (CLASSIFIED) LOCATIONS, CLASSES I, II, AND III, DIVISIONS 1 AND 2

1. b. D; 500.6(A)(4)
2. c. Cl I, Div 1; 500.5(B)(1)(2)
3. c. 3/4"; 500.8(E)
4. b. False; 500.2, definitions
5. d. 40°C; 500.8(C)(4)
6. b. False; 500.7(K)(1)
7. a. True; 500.2, definitions
8. e. none of the above; 500.6(B)(1)
9. b. II and III; 500.6(A), FPN 3
10. c. hermetically sealed; 500.2, definitions

## ⊙ ARTICLE 501–CLASS I LOCATIONS

1. a. True; 501.15(A)(4), Exception 1
2. a. True; 501.10(B)(2)(5)
3. a. True; 501.145
4. d. all choices apply; 501.125(A)(2)– (4)
5. a. True; 501.15(C)(4)
6. c. 3A @ 120V; 501.105(B)(6)
7. b. reducers; 501.15(B)(2)
8. c. Class I, Divisions 1 and 2 locations only; 501.15(C)(2)
9. c. both a and b; 501.10(B)(1)(2)
10. a. True; 501.100(A)(2)
11. c. purging; 501.15(A)(2), FPN 1
12. b. vapors, gases; 501.15, FPN 1
13. b. False; 501.40
14. b. False; 501.10(A)(2)
15. e. 25 percent; c. rigid metal; 501.15(C)(6)
16. a. Class I only; 501.30(A)

17. a. general-purpose; 501.120(B)(2)

18. b. False; 501.15(A)(1)

19. b. 36", 18"; 501.15(A)(3)

20. h. all of the preceding; 501.105(A)

21. b. 5/8"; 501.15(C)(3)

22. b. 18" 501.15(A)(1)

23. a. True; 501.15(B)(2), Exception 4

24. c. threaded rigid metal conduit, threaded steel intermediate conduit; 501.130(A)(3)

25. c. 2"; 501.15(A)(1)(2)

26. b. Class I, Division 1; 501.135(B)(1)

27. a. True; 501.35(B)

28. b. 80; 501.125(B)

29. d. gas/vaportight continuous sheath; 501.15(E)(3), Exception

30. b. threaded; 501.10(A)(1)(a)

31. a. True; 501.15(D)(3)

## ⊙ ARTICLE 502–CLASS II LOCATIONS

1. d. dusttight; 502.120(B)(1)

2. e. none of the preceding; 502.128(A)

3. e. all of the preceding methods; 502.10(A)(2)

4. c. dusttight; 502.10(A)(1)(4)

5. a. True; 502.30(B)

6. b. totally enclosed air-cooled; 502.125(B)

7. b. dust; 502.100(A)(3)

8. a. True; 502.15(2)

9. d. general purpose wireway; 502.10(B)(1)

## ⊙ ARTICLE 503–CLASS III LOCATIONS

1. a. True; 503.130(A)

2. b. noncombustible; 503.160

3. c. 329°F; 503.5

4. a. True; 503.155(B)

5. d. I and IV; 503.10(A) & (B)

## ⊙ ARTICLE 504–INTRINSICALLY SAFE SYSTEMS

1. a. True; 504.30(A)(1)

2. c. light-blue; 504.80(C)

3. b. 2"; 504.30(A)(3)

4. b. intrinsically safe circuit; 504.2, definitions

5. c. 25; 504.80(B)

6. b. junction, terminal; 504.80(A)

7. c. listed; 504.4

## ⊙ ARTICLE 505–CLASS I, ZONE 0, 1, AND 2 LOCATIONS

1. c. identified, listed; 505.9(E)(2)

2. b. maximum experiment safe gap; 505.6, FPN 1

3. a. True; 505.16(B)(5), Exception

4. b. False; 505.5(A)

5. b. False; 505.9(B)(1)

6. b. flameproof "d"; 505.2

7. e. none of the choices apply; 505.18(A)

8. b. Class I, Zone 1; 505.5(B)(2)(2)

9. e. none of the choices apply; 505.16, FPN 2

10. c. combustible gas detection system; 505.2

11. b. False; 505.15(C)(1)(f)

12. c. equipment built to American standards; FPN Figure 505.9(C)(2)

13. d. "n"; 505.2

14. c. 1 3/16"; Table 505.7(D)

## ⦿ Article 506–Zone 20, 21, and 22 Locations for Combustible Dusts or Ignitible Fibers/Flyings

1. b. dust-air; 506.5(B)(3), FPN 3
2. a. True; 506.21
3. b. False; 506.2, Definitions
4. b. False; 506.9(D)
5. c. Zone 21; 506.2, Definitions

## ⦿ Article 510–Hazardous (Classified) Locations–Specific

No questions

## ⦿ Article 511–Commercial Garages, Repair and Storage

1. d. 6"; 511.10(B)(3)
2. b. False; 511.3(C)(3)(b)
3. b. False; 511.10(A)

## ⦿ Article 513–Aircraft Hangars

1. a. True; 513.7(D)
2. a. True; 513.16(B)(2)
3. b. portable equipment; 513.2
4. d. shut off; 513.10(A)(1)
5. c. 5' horizontally, 5'; 513.3(C)(1)
6. a. I only; 513.10(C)(2)

## ⦿ Article 514–Motor Fuel Dispensing Facilities

1. b. Class I, Group D, Division 1; Table 514.3(B)(1)
2. c. Emergency controls; 514.11(C)

3.    c. 100°F; 514.3(A)

4.    b. unclassified; Table 514.3(B)(1)

5.    b. 3', Class I, Division 1; Table 514.3(B)(1)

6.    b. False; 514.9(A)


## ⊙ ARTICLE 515–BULK STORAGE PLANTS

1.    d. Zone 2; Table 515.3

2.    a. True; Table 515.3

3.    b. False; Table 515.3, see (Location) Shell, ends or roof and dike area

4.    a. True; 515.7(A)

5.    c. Class I, Division 1, Zone 1; Table 515.3

6.    b. False; Table 515.3, see (Location) Vent-discharging upward

7.    a. True; 515.8(C)


## ⊙ ARTICLE 516–SPRAY APPLICATION, DIPPING, AND COATING PROCESSES

1.    b. spray area; 516.2, definitions

2.    c. either Class I or Class II, Division 1; 516.3(B)(3)

3.    b. False; 516.3(C)(6)

4.    a. True; 516.10(A)(1), Exception

5.    b. identified; 516.10(C)(4)

6.    b. grounded; 516.10(A)(6)


## ⊙ ARTICLE 517–HEALTH CARE FACILITIES

1.    b. False; 517.61(C)(2)

2.    b. False; 517.18(A)

3.    c. X-ray installations, momentary rating; 517.2

4.    b. 150kVA; 517.34, Exception

5.    b. wet location; 517.2

6.  a. True; 517.80, FPN

7.  d. 4; 517.2

8.  a. 4; 517.18(B)

9.  c. Class I, Division 1; 517.60(A)(2)

10. b. total hazard current; 517.2

11. e. none of the preceding; 517.13(B)

12. b. 60A; 517.72(A)

    LT Rating (@50 percent) = 84A x .5 = 42A (greater of two)

    ST Rating (@100 percent) = 35A

    Minimum size standing disconnecting means must be rated for 60A.

13. b. False; 517.160(B)(1)

14. a. True; 517.33(A)(9)

15. b. False; 517.2, Patient Care Area, FPN

16. a. True; 517.43(B), Exception

17. d. 50 percent, 100 percent; 517.73(A)(1)

18. c. Nursing homes; 517.40(C)

19. c. #10 AWG; 517.14

20. a. True; 517.2

21. b. False; 517.21

22. c. transformer; 517.160(A)(4)

23. c. readily; 517.41(E)

24. b. X-ray installations, long-time rating; 517.2

25. c. 60A; 517.71(B)

26. d. all of the preceding; 517.30(D)


## ⊙ ARTICLE 518–ASSEMBLY OCCUPANCIES

1.  a. True; 518.4(B)

2.  b. False; 518.5

3.  b. 100; 518.1

# ⊙ ARTICLE 520–THEATERS, AUDIENCE AREAS OF MOTION PICTURE AND TELEVISION STUDIOS, PERFORMANCE AREAS, AND SIMILAR LOCATIONS

1. c. 20A; 520.41(A)
2. b. 150V; 520.25(C)
3. b. 3.3'; 520.69(C)
4. d. bundled; 520.2, definitions
5. d. 100 feet; 520.53(J)
6. d. listed; 520.48
7. b. False; 520.2, definitions

# ⊙ ARTICLE 522–CONTROL SYSTEMS FOR PERMANENT AMUSEMENT ATTRACTIONS

1. a. 300V; 522.10(B)
2. b. False; 522.24(B)(3)(1)
3. a. True; 522.2, definitions
4. e. answer not available; 522.21(B), Exception
5. b. False; 522.7
6. a. True; Table 522.22, Note 2

# ⊙ ARTICLE 525–CARNIVALS, CIRCUSES, FAIRS, AND SIMILAR EVENTS

1. b. bonded; 525.30
2. d. weatherproof; 525.22(A)
3. c. 2 AWG; 525.20(B)
4. a. True; 525.5(B)(1)

# ⊙ ARTICLE 530–MOTION PICTURE AND TELEVISION STUDIOS AND SIMILAR LOCATIONS

1. c. 20A; 530.12(B)
2. c. 250VDC; 530.64(A)
3. c. bull switch; 530.2

4.  d. 400 percent; 530.18(B)

5.  a. True; 530.2

6.  b. False; 530.23

## ⊙ ARTICLE 540–MOTION PICTURE PROJECTION ROOMS

1.  b. listed; 540.32

2.  d. No. 8 AWG; 540.13

3.  a. True; 540.11(B), Exception 2

4.  a. True; 540.2

## ⊙ ARTICLE 545–MANUFACTURED BUILDINGS

1.  a. True; 545.6, Exception

2.  d. any of the preceding choices; 545.12

## ⊙ ARTICLE 547–AGRICULTURAL BUILDINGS

1.  c. Slatted; 547.10(B)

2.  b. False; 547.10

3.  d. all of the preceding choices; 547.5(G)

4.  a. True; 547.2, distribution point, FPN 1

## ⊙ ARTICLE 550–MOBILE HOMES, MANUFACTURED HOMES, AND MOBILE HOME PARKS

1.  a. True; 550.18(B)(2)

2.  b. False; 550.13(B)

3.  c. 30 ft.; 550.32(A)

4.  c. 20'; 550.10(D)

5.  c. No. 8 copper; 550.16(C)(2)

6.  b. False; 550.2, definitions

7.  c. grounded; 550.16

8.  b. 3 VA/ft²; 550.12(A)

9.  b. increase to 16,000 volt-amperes; 550.31(1)

10. b. built-in microwave; 550.2, definitions (appliance, portable, FPN)

11. b. False; 550.33(B)

12. b. 1-minute, 900-volt; 550.17(A)

13. c. 40A; 550.10(A), Exception 1

14. d. 8.4kW; 550.18(B)(5)


## ⊙ ARTICLE 551–RECREATIONAL VEHICLES AND RECREATIONAL VEHICLE PARKS

1.  d. 10A; 551.51(A)(1)

2.  b. False; 551.20(B)

3.  b. False; 551.76(C)

4.  a. True; 551.56(E)

5.  d. answer not provided; 551.77(D)

6.  c. locking system; 551.45(B), Exception 2

7.  b. False; 551.73(A)

8.  c. 6; 551.42(D)

9.  b. procedure for inserting or removing plug; 551.77(F)

10. b. camping trailer; 551.2

11. b. False; 551.31(D)

12. a. weatherproof; 551.78(A)

13. d. 50kVA; Table 551.73

    3600VA x 33 = 118,800VA or 118.8kVA

       33 RV sites = 42 percent (.42) demand factor (Per Table 551.73)

    118.8kVA x .42 = 49.896kVA (approximately 50kVA)

14. a. True; 551.43(C)

15. b. 240V; 551.72

16. e. all of the preceding; 551.30(E)

17. a. True; 551.41(C)(4), Exception

18. b. 3" x 1.75"; 551.46(D)

19. b. False; 551.71

## ⊙ ARTICLE 552–PARK TRAILERS

1. b. False; 552.10(E)(1)
2. a. True; 552.47(B)(2)
3. c. 400 ft²; 552.2, definition
4. a. 2'; 552.41(D)(1)
5. e. answer not provided; 552.46(B)(3)(c)
6. b. readily accessible; 552.45(B)
7. c. secured; 552.60(B)
8. a. True; 552.41(F)(1)
9. c. 7.76kW; 552.47(B)(5)
10. b. False; 552.56(B)
11. b. 1; 552.46(B)(1)


## ⊙ ARTICLE 553–FLOATING BUILDINGS

1. b. insulated; 553.9
2. a. True; 553.4


## ⊙ ARTICLE 555–MARINAS AND BOATYARDS

1. d. larger; Table 555.12, Note 1
2. b. False; 555.19(A)(4)
3. d. 600; 555.4
4. d. all of the preceding; 555.13(B)(1)
5. b. False; 555.9


## ⊙ ARTICLE 590–TEMPORARY INSTALLATIONS

1. c. continuity; 590.6(B)(2)(a)(1)
2. b. False; 590.4(D)
3. d. 90 days; 590.3(B)

## ARTICLE 600–ELECTRIC SIGNS AND OUTLINE LIGHTING

1. a. True; 600.32(G)
2. c. 14'; 600.9(A)
3. a. True; 600.42(F)
4. b. externally; 600.6
5. c. 50'; 600.32(J)(2)
6. d. 1000V; 600.32(I)
7. a. True; 600.5(A)
8. c. 300 mA; 600.23(D)
9. b. False; 600.23(B)(1)
10. a. No. 14; 600.7(a)

## ARTICLE 604–MANUFACTURED WIRING SYSTEMS

1. c. polarized; 604.6(C)
2. a. True; 604.4, Exception 2

## ARTICLE 605–OFFICE FURNISHINGS (CONSISTING OF LIGHTING ACCESSORIES AND WIRED PARTITIONS)

1. c. 13; 605.8(C)
2. c. 9'; 605.5(B)
3. a. True; 605.2, Exception

## ARTICLE 610–CRANES AND HOISTS

1. b. separate; 610.42(B)(3)
2. c. one; 610.15
3. b. False; 610.55
4. b. .84; Table 610.14(E)
5. c. 50 percent, 75 percent; 610.33

6.  b. No. 20 AWG; 610.14(C)(2)

7.  b. False; 610.21(H)

8.  b. False; 610.14(G)

9.  d. both I and II; 610.51(B)

10. b. False; 610.11(A)

11. c. 7.5HP; 610.43(D)

12. c. 1/3; 610.42(B)(1)

13. c. 4 Table; 610.14(D)

## ⊙ ARTICLE 620–ELEVATORS, DUMBWAITERS, ESCALATORS, MOVING WALKS, PLATFORM LIFTS, AND STAIRWAY CHAIR LIFTS

1.  c. duplex; 620.23(C)

2.  b. ground-fault circuit interrupter; 620.85

3.  a. True; 620.3(C)

4.  d. 15'; 630.32

5.  b. No. 20 AWG copper; 620.12(A)(1)

6.  a. True; 620.41(3), FPN

## ⊙ ARTICLE 625–ELECTRIC VEHICLE CHARGING SYSTEM

1.  c. 25'; 625.17

2.  a. 1066; Table 625.29(D)(2)

3.  d. 60, 150; 625.23

4.  b. False; 625.21

## ⊙ ARTICLE 626–ELECTRIFIED TRUCK PARKING SPACES (NEW)

1.  d. 25' 626.25(B)(3)

2.  c. 2', 6.5' 626.22(C)

3.  a. True; 626.11(D)

4.  d. 208/120V, 3ϕ, 4-wire system; 626.10

5.   a. True; 626.32(A)

6.   d. 25'; 626.25(B)(3)

7.   b. False; 626.2, definition–Electrified Truck Parking Space, FPN

8.   h. I and II or III; 626.24(B)(1) & (2)

9.   c. 39 percent; Table 626.11(B)

## ⊙ ARTICLE 630–ELECTRIC WELDERS

1.   c. actual primary current; 630.31(B), FPN 2

2.   a. True; 630.42(C)

3.   a. True; 630.11(B), Exception, FPN

4.   b. flame retardant; 630.41

5.   b. equipment grounding; FPN to 630.15

## ⊙ ARTICLE 640–AUDIO SIGNAL PROCESSING, AMPLIFICATION, AND REPRODUCTION EQUIPMENT

1.   d. input or output; 640.9(D)

2.   c. tripping; 640.45

3.   d. No. 14 copper; 640.7(A)

4.   b. False; 640.2, Definitions

5.   c. Class 1; 640.9(C)

6.   b. False; 640.2, Definitions

## ⊙ ARTICLE 645–INFORMATION TECHNOLOGY EQUIPMENT

1.   d. both a and b; 645.10

2.   a. True; 645.5(D)(1)

3.   d. all of the preceding; 645.16

4.   b. 125 percent 645.5(A)

## ⊙ ARTICLE 647–SENSITIVE ELECTRONIC EQUIPMENT

1. b. False; 647.4(D)(2)
2. a. True; 647.8(C)
3. b. False; 647.6(B)
4. b. system voltage; 647.4(B)

## ⊙ ARTICLE 650–PIPE ORGANS

1. c. No. 14; 650.6(A)
2. d. 6A; 650.8
3. b. 30VDC 650.4

## ⊙ ARTICLE 660–X-RAY EQUIPMENT

1. b. False; 660.48, Exception
2. b. approved; 660.10
3. c. 30 amperes; 660.4(A)

## ⊙ ARTICLE 665–INDUCTION AND DIELECTRIC HEATING EQUIPMENT

1. d. 50V, 5 minutes 665.24(460.28)
2. b. False; 665.7(B)
3. e. none of the preceding; 665.2

## ⊙ ARTICLE 668–ELECTROLYTIC CELLS

1. d. 600; 668.21(A)
2. c. gases; 668.40
3. a. True; 668.10(B)
4. c. insulated; 668.32(A)
5. b. False; 668.14(A)

## ● ARTICLE 669–ELECTROPLATING

1. a. True; 669.1
2. b. False; 669.8(B)

## ● ARTICLE 670–INDUSTRIAL MACHINERY

1. d. either a or b; 670.4(C)
2. a. True; 670.4(B)

## ● ARTICLE 675–ELECTRICALLY DRIVEN OR CONTROLLED IRRIGATION MACHINES

1. b. False; 675.10(A)(1)
2. b. Table 250.122; 675.13
3. b. False; 675.2, Definitions
4. d. all of the preceding; 675.12
5. c. 167°F; 675.4(A)

## ● ARTICLE 680–SWIMMING POOLS, FOUNTAINS, AND SIMILAR INSTALLATIONS

1. b. False; 680.44(B)
2. d. 18"; 680.23(A)(5)
3. b. 6'; 680.22(A)(2)
4. b. False; 680.27(B)(2)
5. d. 6'; 680.23(F)(1), Exception
6. a. True; 680.73
7. c. drained; 680.2, definitions
8. b. False; 680.23(B)(2)
9. c. either a or b; 680.43(B)(1)(a) & (b)
10. a. drinking fountain; 680.2, definitions
11. b. False; 680.22(B)
12. d. 300V; 680.51(B)

13. c. 125 percent; 680.9

14. d. choices a–c; 680.26(C)

15. b. False; 680.6(2)

16. b. 30A; 680.43(A)(2)

17. c. therapeutic tubs; 680.62(B)(5)

18. b. shall be mounted at least 12 feet vertically above the pool deck; 680.27(C)(2)

19. b. False; 680.2, definitions

20. c. both a and b; 680.23(B)(3)

21. b. False; 680.57(C)(2)

22. a. True; 680.42(A)(2)

23. d. correct choice not given; 680.21(A)(1)


## ⊙ ARTICLE 682–NATURAL AND ARTIFICIALLY MADE BODIES OF WATER

1. c. shoreline; 682.2, definitions

2. a. True; 682.33(B)

3. b. False; 682.31(A)

4. a. True; 682.11


## ⊙ ARTICLE 685–INTEGRATED ELECTRICAL SYSTEMS

1. b. False; 685.12

2. c. unitized; 685.1


## ⊙ ARTICLE 690–SOLAR PHOTOVOLTAIC SYSTEMS

1. a. True; 690.56(B)

2. a. True; 690.8(B)

3. d. all of the preceding; 690.4(C)

4. b. solar cell; 690.2, definitions

5. d. No. 2/0 AWG; 690.74

6. d. 50V; 690.71(B)(1)

7. b. False; 690.2, definitions

8. d. USE-2; 690.31(B)

9. b. photovoltaic power source; 690.9(B), Exception

## ⊙ ARTICLE 692–FUEL CELL SYSTEMS

1. b. False; 692.61, FPN

2. c. stand-alone system; 692.2, definitions

3. b. primary; 692.54

4. b. False; 692.2, definitions

## ⊙ ARTICLE 695–FIRE PUMPS

1. b. False; 695.6(E)

2. b. 1-hour; 695.14(F)

3. c. either a or b; 695.3(A)(1)

4. b. False; 695.6(F)

5. a. True; 695.4(B)(2)

6. b. 5 percent; 695.7

## ⊙ ARTICLE 700–EMERGENCY SYSTEMS

1. d. 2-hour; 700.12(B)(2)

2. d. 1000; 700.9(D)(2)

3. e. all of the preceding choices; 700.7(A)– (D)

4. d. all of the preceding choices; 700.6(A)

5. d. all of the preceding; 700.21

6. a. True; 700.20

7. a. True; 700.4(B)

8. d. 15-minute; 700.12(B)(1)

9.   e. none of the preceding; 700.12(A)

10.  a. True; 700.22


## ⊙ ARTICLE 701–LEGALLY REQUIRED STANDBY SYSTEMS

1.   a. True; 701.15

2.   b. at the service entrance; 701.9(A)

3.   b. operating rooms; 701.2, FPN

4.   a. True; 701.11(A)


## ⊙ ARTICLE 702–OPTIONAL STANDBY SYSTEM

1.   b. optional system; 702.2

2.   b. False; 702.5(B)(1)


## ⊙ ARTICLE 705–INTERCONNECTED ELECTRIC POWER PRODUCTION SOURCES

1.   d. all of the preceding; 705.14

2.   c. parallel; 705.143

3.   b. False; 705.10, Exception

4.   d. 1000V; 705.21

5.   e. none of the preceding; 705.12(A)

6.   b. False; 705.32


## ⊙ ARTICLE 708–CRITICAL OPERATIONS POWER SYSTEMS (COPS) (NEW)

1.   c. above; 708.11(B)

2.   b. False; 708.52(D)

3.   a. True; 708.6(D)

4.   d. designated critical operations area; 708.2, definitions

5.   c. +/- 10 percent; 708.22(C)

6.   e. answer not provided; 708.20(F)(1)

7.   b. False; 708.4(A)

## ⊙ Article 720–Circuits and Equipment Operating at Less Than 50 Volts

1. b. False; 720.7
2. c. 660W; 720.5
3. b. False; 720.4

## ⊙ Article 725–Class 1, Class 2, and Class 3 Remote-Control, Signaling, and Power-Limited Circuits

1. d. all of the preceding; 725.133(H)
2. d. only I and II; 725.49(B)
3. e. all of the preceding; 725.179(E)
4. b. 5 seconds; 725.41(A)(2)
5. c. 30 volts or more, No. 6; 725.121(A)(5)
6. c. either a or b; 725.3(C), Exception
7. a. True; 725.179(L), FPN
8. a. True; 725.48(B)(1)
9. b. PTLC; 725.154(D)(1)
10. b. Class 2; 725.2, definitions
11. a. True; 725.143
12. d. 12" ; 725.127, Exception
13. a. True; 725.31(B)
14. a. True; 725.51(A)

## ⊙ Article 727–Instrumentation Tray Cable: Type ITC

1. a. True; 727.8
2. b. 300V; 727.6
3. b. False; 727.1
4. c. damage; 727.10

# ● ARTICLE 760–FIRE ALARM SYSTEMS

1. c. 7'; 760.53(A)(2)

2. c. 30 percent; 760.179(H

3. e. none of the preceding; 760.43

4. a. True; 760.1, FPN No.1

5. a. True; 760.142

6. d. 20A; 760.127

7. c. either a or b; 760.2, definitions


# ● ARTICLE 770–OPTICAL FIBER CABLES AND RACEWAYS

1. a. True; 770.100(A)(1)–(3)

2. e. all of the preceding; 770.6

3. b. 4'11"; 770.179(C), FPN

4. b. False; 770.2


# ● ARTICLE 800–COMMUNICATION CIRCUITS

1. c. No. 12; 800.106(B)

2. c. 6'; 800.53

3. b. 2" ; Table 800.133(A)(2)

4. c. demarcation; 800.2, definitions

5. a. True; 800.133(A)(1)(b), Exception

6. c. CMG; Table 800.154(E)

7. b. False; 800.2, definitions

8. c. one-and two-family dwellings; 800.100(A)(4)

9. b. False; 800.90(B)

10. c. both I and II; 800.44(A)(2) & 800.44(B), Exception No. 3

# ⊙ Article 810–Radio and Television Equipment

1. a. True; 810.21(E)

2. c. 350V; 810.71(C)

3. c. 17'; Table 810.16(A)

4. e. correct choice not provided; 810.11, Exception

5. d. 14'; Table 810.52

6. c. No. 17 AWG; 810.21(H)

7. a. True; 810.13

# ⊙ Article 820–Community Antenna Television and Radio Distribution Systems

1. c. 6'; 820.44(F)(3)

2. c. hotels; 820.154(C)(5)

3. b. 60V; 820.15

4. a. True; 820.93, FPN

5. a. True; 820.44(C)

6. d. CATVR; Table 820.154(E)

# ⊙ Article 830–Network-Powered Broadband Communications Systems

1. b. 11.5'-; 830.44(D)(2)

2. c. I and III; Table 830.154

3. a. True; 830.2, definitions

4. d. 0 Table; 830.47

5. b. 1 minute; Table 830.15, Note 1

6. a. True; 830.100(B)(3)(2)

1.  a. True; Table 5

2.  b. 55.80 mm$^2$; Table 8

3.  b. 5 sec; Note 1 to Table 11B

4.  d. 24" ; Note 4 to Tables

5.  a. True; Table 1, FPN No. 2

6.  b. .7250; Table 5A

7.  c. .20Ω; Table 9

8.  d. 8.179 in$^2$; Table 4

9.  c. Table 12(B)

10. d. 1.320Ω/km; Table 8

11. b. False; Table 1

12. d. 4110; Table 8

13. b. False; Note 5 to Tables

14. c. 500,000; Table 8

15. b. .9923; Table 5

16. b. .040Ω; Table 9

17. c. Table 8; Note 8 to Tables

18. c. 16" ; Table 2 (One Shot & Full Shoe Benders)

19. b. False; (.182") Table 5

20. b. False; Table 12(A), Heading

21. a. .051Ω; Table 9

22. a. True; Table 5

23. d. 12.127 in$^2$; Table 4

24. b. C; Note 1 to Tables

25. c. .1901 in$^2$; Table 5

26. a. .778Ω/kFt. ; Table 8

27. b. .327 in$^2$; Table 4

28. a. .0730; Table 5A

29. d. 5A; Table 11A

30. a. True; Note 3 to Tables

31. c. 10,380; Table 8

32. e. choice not available; (2.693 in²) Table 4

33. b. 9 1/2"; Table 2


## ⦿ ANNEX B–APPLICATION INFORMATION FOR AMPACITY CALCULATION

1. c. 19.84A; Table B.310.11 & Formula

$$\frac{\sqrt{(.5)(35)}}{10} \times 25A \times .60 = 19.84A$$

2. b. False; Note to Table B.310.8

3. d. 25A; Table B.310.1


## ⦿ ANNEX C–CONDUIT AND TUBING FILL TABLES FOR CONDUCTORS AND FIXTURE WIRES OF THE SAME SIZE

1. a. True; see Note 1 to Tables C.1–C.12

2. a. True; see footnote to Tables C.1(A)–C.12(A)

3. d. 81; Table C.1

4. d. no choices apply; 4"- Type EB, PVC Table C.12

5. b. False; see Notes to Tables C.1–C.12

# ABOUT THE AUTHOR

During the past eleven years, Alvin Walker, a native of Shreveport, Louisiana, has owned and operated a small yet successful electrical contracting business. In his over twenty-five years of experience, he has developed a very strong background in electrical maintenance, design, and construction applications. He has taught the National Electrical Code, Electrical Theory, Business and Law for Contractors, and other basic and advanced electrical classes at Louisiana State University–Shreveport, Bossier Parish Community College, Louisiana Technical College, and Houston Community College–Stafford. Mr. Walker is best known for his hands-on approach and the ability to simplify and explain the most difficult electrical subject matters. He is a master electrician and holds a Louisiana state license as an electrical contractor. He has a degree in electrical engineering from the University of South Carolina and has worked as an electrical engineer for E.I. DuPont, Westinghouse, and M.W. Kellogg. He is currently in the process of publishing an electrical theory book and a volume of books extensively covering the National Electrical Code. As an author, his primary objective is to produce material that is thoroughly explained, easily understood, and user friendly.